Adventures in Recreational Mathematics
A Series of Excursions into the Realms
of the Wondrous Abstract

Isky Mathews

D1318579

Preface

My momma always used to say:
Life is like a box of chocolates...
You never know what you're
gonna get.

Forrest Gump

Mathematics is, for those who know and love it, like a delicious cake with a hard crust. You need to know quite a bit to get to "the good stuff" but once you get there, it can be more remarkable than anything possible in this world or any other.

Johannes Gutenberg introduced printing to Europe, thereby opening the way for much wider circulation of books and knowledge. This huge increase in the depth and breadth of knowledge gave birth to the Renaissance and revolutionary breakthroughs in science.

Up until a decade or so ago, most advanced, specialist texts on mathematics, physics and other sciences could only be accessed in university libraries. A motivated individual of pre-university age could find "popular science" books on the high street but nothing at degree level, much less research papers. In the last few years this has changed -the birth of the internet, invention of efficient search engines and their global reach combine to represent "Gutenberg-squared". School children from all over the planet can now read manuscripts of almost any level up to and including cutting edge research. I believe this massive increase in availability will lead to a series of even greater leaps forward in scientific and mathematical thought than ever seen before. Previously only a few tens of thousands of university students or even fewer post graduates were introduced to these advanced topics, nowadays billions of people have these opportunities and even if only 0.01% take advantage of it, that will still involve at least a 10 fold increase, putting them in a position to make further breakthroughs.

However, even in this age of the internet, one of the obstacles to making really significant progress is "learning without a teacher". There's no one there to say "if you want to understand that material, it's a good idea to read this one first" or to lend a helping hand with the great onslaught of definitions, notation and terminology thrown at you. Another issue concerns finding your "sweet spot". You can easily source classical, jazz, swing, rock and various genres of pop music before deciding which is to your liking. As a beginner in mathematics, you won't have even heard of many of the most interesting ideas, theorems, or fields and it is not at all obvious how to uncover them - you may, after struggling around the literature, decide that number theory is most to your liking whereas given the right explanations, you would have actually preferred geometry. This book aims help you to overcome both of these issues.

The book is built from a collection of articles I wrote for a school magazine between years 9 and 11, reflects some of the mathematics that I've gleaned from just these sorts of struggles and my hope is that it will help to surmount both of these problems.

Each chapter aims to introduce completely different areas of mathematics or ideas for you to play around with, starting with simple concepts and later building up to some of the results at the frontiers of our knowledge on a subject. As a result, each text aims to be completely self contained but there are some minimal pre-requisites:

- It would be useful what sets, subsets and functions are. You should know that $x \in S$ means that the object x is in the set S, that $X \subset S$ means that the set X is a strict subset of S and that $X \subseteq S$ means that X is a subset of S and that it could be equivalent to S. You should also know what the terms "union" and "intersection" mean for sets.

- You should know that $\mathbb{C}$ is the set of complex numbers, that $\mathbb{R}$ is the set of *real numbers*, that $\mathbb{Q}$ is the set of *rational numbers* (the "Q" stands for *quotient*, which means "fraction"), that $\mathbb{Z}$ means the integers and that $\mathbb{N}$ means the set of nonnegative integers.

You shouldn't worry too much about this - anything else is sufficiently obscure that I will give a definition or some explanation for it!

I hope you enjoy reading the book as much as I enjoyed writing it.

Acknowledgements

Thank you to Wikimedia Commons and their creators who place such useful images in the public domain. I wish to give attribution to Robert Webb's Stella software, which produced the image of the Great Dirhombicosidodecahedron on the front, which can be found at www.software3d.com/Stella.php.

Thank you for my friend Aldwin Li, who illustrated some of the chapter covers and was a general aesthetic guide. Thank you to my friend Joshua Loo, who diligently edited the original publication from which these articles originate and gave me great support in both motivation and typesetting. Thank you to all the authors of any mathematical book, blog or paper I've read who have shown me all these wondrous things.

I wish to finally thank my father for his physics enthusiasm that got me started in my mathematical readings and his great push to make me publish.

Chapter 1

Tilings

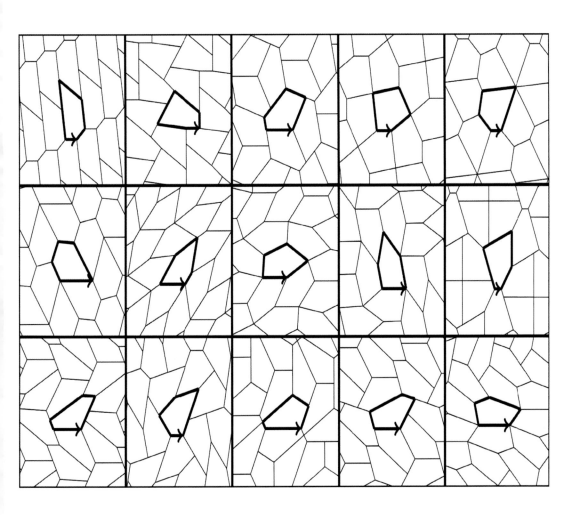

Salutations!

Mathematics can be confusing, it can be surprising but it can also be a lot of fun; this is what *Adventures in Recreational Mathematics* (ARM) aims to tap into!

We are used to seeing physical tiles covering a bathroom wall, the floor of a public buildings or the pavements we walk over. They also occur in nature, for example the cells of a beehive, the surfaces of crystals, the cracks in a dried up river bed. This article is going to investigate the many possible shapes and properties of tiles that can be used to fill a flat surface leaving no gaps. Restating this in more technical language, which collections of polygons can tile the plane? Are there nice geometric principles that can allow us to decide this matter and can we categorise tilings into different types? As we will see, not only do these questions lead to some interesting puzzles but they allow us to see great beauty in patterns both visual and mathematical.

Tilings of the plane will, quite obviously, involve infinite numbers of tiles (as they have to go on forever in every direction) but those that people often examine use only a finite number of different shapes in these tilings (e.g. quadrilaterals, triangles etc. with particular side lengths and angles) - each such shape we will call a *prototile* of a given tiling. In a sense, you can imagine the set of prototiles to be the "parts" that a construction worker hoping to make a tiling for you might put on his shopping list (but he would still need an infinite number of each to make it).

We begin, in Fig. 1.1, with the simplest: *regular* tilings, the only three of which are shown in Fig 1.1.

To introduce the bountiful quantities of terminology in the field, we can make a number of observations about these tilings:

1. A given edge of a tile in each tiling touches no more than one edge from another tile - tilings with this property are called *edge-to-edge*.

2. Every vertex has an identical arrangment of tiles around it, so we call them *archimedean*.

3. For every 2 vertices, there is some combination of rotations and translations that switches the location of the two vertices and retains the tiling's overall shape (i.e. where there was a line segment, there is now a line segment; where there was a hexagon, there is now a hexagon etc.) - tilings with this property are called *uniform*. In other tilings, it may not be true that *every* two vertices can be "exchanged" in this manner but that instead the vertices can be assigned types, where every vertex of a given type can be "exchanged" with every other vertex of the same type - tilings with k types are then called *k-uniform*.

4. Each prototile in each tiling is a regular polygon.

5. There is only one prototile.

6. Every tile has unit side length.

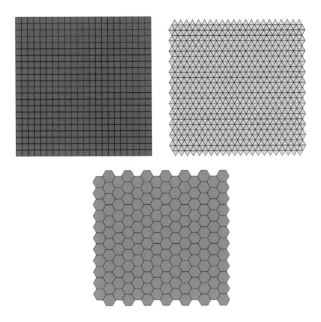

Figure 1.1: Three familiar tilings.

Indeed, for a tiling to be *regular*, it must have the $1^{st}, 4^{th}, 5^{th}$ and 6th properties we mentioned and, if we negate the property of there only being one prototile, we get *semi-regular* tilings (some lovely examples of which are shown in Figure 1.2). As practise, can you tell which properties out of those mentioned in the bullet-points above each tiling has?

You may have also recognised the hexagonal tiling to be that often used by honeybees in the construction of their hives. It is commonly believed that the reason they evolved to do this is that hexagons cover the most area while minimising the total sidelengths of the tiles used or, in their case, minimising the amount of wax used[1]

Perhaps unsurprisingly, it is trivial to prove that for $\forall n \in \mathbb{R}, n > 2$, there is an irregular n-gon which tessellates. This can be seen by constructing the shapes in Figure 1.2.

The top-left part of the figure displays a tessellating hexagon and by adding another pair of sides on top, we get the adjacent polygon (a tessellating octagon). We can continue adding pairs like this to get tessellating $2n$-gons for every n. The third part of the image shows how one could adjust the flat top-sides of these $2n$-gons to create "forked tongue" endings which mesh together such that the shape still tessellates and now has $2n + 1$ sides. Thus, one can construct all even and odd numbered n-gons such that they tile for $n > 4$ using this method!

---

[1]The fact that hexagons were better in this respect than other single-polygon tilings was proved in antiquity but proving it was the best in comparison to *all* other possible tilings resisted proof until as recently as 1998!

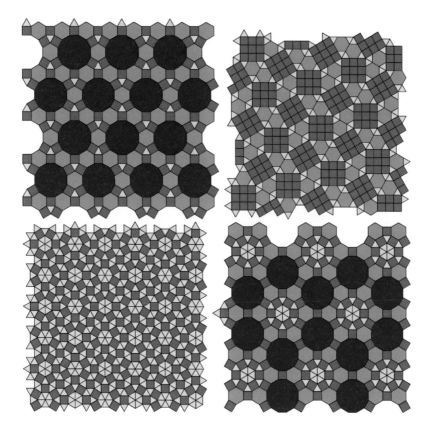

Figure 1.2: Pleasing patterns and aesthetic arrangements!

Let's call a tile *convex* if all of its internal angles are $< 180°$. Then contrary to the generality implied by the above result, I can tell you that there is no tiling of the plane with convex tiles of 7 sides or more. Famously, the sum of the exterior angles of an n-gon must sum to $360°$ (which you can see from the fact that by moving along the polygon, you must have turned a full circle by the time you reach your starting point) and calling the ith exterior angle E_i & thus the ith interior angle $(180° - E_i)$ we have that

$$\sum_i (180 - E_i) = 180n - \sum_i E_i = 180n - 180 \times 2 = 180(n - 2)$$

Then consider the fact that in a tiling of the plane by convex tiles at least 3 tiles must meet at each point and so at such a point, the average angle must be no more than $360/3 = 120°$. But in an n-gon where $n > 6$, the average angle is (by what we said above) $\frac{180(n-2)}{n} = 180(1 - \frac{2}{n}) > 180(1 - \frac{2}{6}) = 120°$ showing that it violates our tiling condition and thus cannot tile the plane.

There are, as you can imagine, many other ways of constructing tessellating n-gons than the way illustrated in Figure 1.3 and in fact many ways of tiling

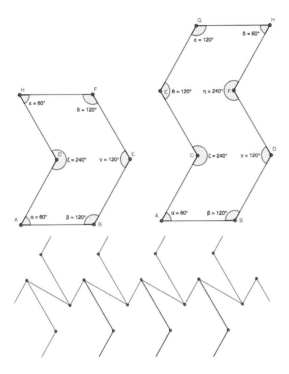

Figure 1.3: Non-convex generality.

them. If we allow ourselves to tile on a plane with negative curvature (commonly known as the *hyperbolic plane*) then you can have significantly more tilings – for example, in the Euclidean plane we tile 6 equilateral triangles around a point but in the hyperbolic plane you can have 7, 8, 9 all the way to an infinite number of triangles around a point! This geometry requires greater examination on its own but that is not the focus of the article[2].

These have all been quite simple examples and concepts for a reader such as yourself! There are numerous significantly complicated ideas in the field, some of which we will now examine.

A tiling problem is one in which you are given a set of prototiles and you have to decide whether it's possible to use all of these prototiles in an arrangement that covers the whole plane (i.e. whether they, well, *tile!*). Mathematicians often then used to refer to "the" tiling problem, which is to create a set of methodical rules to follow by which you could guarantee that using them you could solve *any* tiling problem. This generality is important - most people can

---

[2]This is discussed in greater depth in Chapter 6 of this book.

see quickly that equilateral triangles could tile the plane and also that regular heptagons cannot (so they can clearly solve *some* tiling problems) but could they describe a procedure that they could use that would work on all conceivable tiling problems? A subset of this problem was first proposed for study by the philosopher and mathematician HAO WANG in 1961, after discovering sets of aperiodic tiles called *Wang tiles.*

A periodic tiling is one for which one can pick up the entire plane, translate it by some distance, place it down again and end up with the same tiling. An *aperiodic* tiling, then, is one that does not have any form of periodicity, i.e. for which there exists sections of tiles with the property that if you were to pick them up and move them around the plane without any rotations or scalings you would never find any other tiles of both the same size *and* orientation - these, on their own, are not particularly interesting as, in some sense, *most* tilings are aperiodic (e.g. the hexagonal tiling from above, except that one and only one hexagon is split down the middle of two opposite sides into 2 tiles). What are *extremely* interesting are sets of prototiles with the property that you can *only create aperiodic tilings with them* (which we shall call an aperiodic set). For many years, it was not known if such a thing could possibly exist until Wang explicitly constructed an aperiodic set.

The colours on each side of a Wang tile are a visual way of representing rules for placing them next to each other – here only same-coloured edges can touch but the same rules could just as easily be shown as indents and projections on the edges of each tile. Due to these "matching rules", these tiles have become colloquially known as Wang dominoes and the problem of determining whether a set of Wang tiles can tile the plane is often called the Domino Problem - an example of such a tiling is shown in Figure 1.4.

In 1966, Wang's student ROBERT BERGER showed that the Domino Problem is actually *undecidable* (and thus, by extension, "the" tiling problem in general is too). This means that there is *no* algorithm or procedure you can create that can solve every instance of a tiling problem involving Wang tiles, even if you had a computer with infinite computational capabilities and infinite storage capacity. The details of this proof we will leave to another ARM but, in short, Berger did this by assuming that such an algorithm *did* exist and then showed that using this he could construct another algorithm that could solve the *Halting Problem*, the first ever problem shown to be undecidable (by ALAN TURING, inventor of computability and master of cryptanalysis). Undecidability as a property is quite scary and it often suggests that a problem can get, in some sense, arbitrarily complicated - Wang tiles and the property of aperiodicity are therefore extremely tricky things...

Recall, however, that although "the" tiling problem is undecidable, it doesn't mean that all individual instances of tiling problems are impossible to solve, and so a natural question is to ask whether there are collections of prototiles defined by one or more properties such that if we restrict our tiling problems to come from them, "the" tiling problem is decidable.

The mathematician HEINRICH HEESCH was interested in whether this was possible in situations where we restrict our attention to tilings with a single

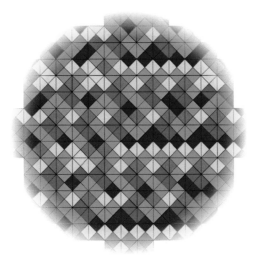

Figure 1.4: A chaotic colouring.

prototile and noticed that even if a prototile couldn't tile the plane, different tiles could cover different maximum areas before no more could be placed without leaving gaps, e.g. the Heesch number of the pentagon is 1 and any quadrilateral's Heesch number is ∞ (as any quadrilateral can tile the plane). The idea of the *Heesch number* of a tile, which is the maximum number of layers of copies of a given prototile can completely surround the previous layer's perimeter, was therefore devised as a natural quantitative indicator of the extent to which a tile "tiles well" on the plane and the *Heesch Problem* is to determine which finite natural numbers are the Heesch number of some tile.

Can we have a tile of finite Heesch number > 2 ? It appears so: in 1995, ROBERT AMMANN found a tile of Heesch number 3 and also constructed a tile, using indentations and projections on sides of hexagons, of Heesch number 4 and subsequently CASEY MANN found an infinite family of tiles of Heesch number 5, which is the highest finite number known to date (see Figure 1.5).

Can you see why Ammann's tiling cannot be expanded beyond 5 layers?

One of the goals of the Heesch problem is to see whether there is a maximum finite Heesch number. If this was true then you could construct a finite tiling-problem algorithm for single prototile tilings: simply attempt to surround the prototile by itself m times and if you can clearly continue then its Heesch number must be infinite, thus proving that it tessellates. This algorithm wouldn't be particularly efficient (at minimum, it would be an EXPTIME, potentially even EXPTIME-complete, algorithm for those who are of the computational complexity persuasion) but it would represent a great step forward.

Figure 1.5: *Left*, a tile of Ammann with Heesch number 4, *right*, an example from Mann's infinite family of tiles with Heesch number 5.

⎯⎯⎯⎯⎯⎯⎯⎯⎯⎯⎯⎯⎯⎯⎯⎯⎯⎯⎯⎯⎯⎯⎯⎯⎯⎯⎯⎯⎯⎯⎯

Given what we wrote above in earlier sections of this article, we know that there's no single convex-prototile tiling with the prototile having 7 or more sides so a natural (but extremely difficult) question given such a bound is to ask *exactly which* convex polygons can tile the plane. Parts of this question are simple: every triangle and every quadrilateral can tile the plane[3]. The classes of convex hexagons that could tile the plane have been known for many years but the final part of classifying the convex pentagons that tile the plane was an open problem up until just late 2017[4]. To understand something of this result, we must first explain what exactly a "class" of pentagons entails.

Suppose we have our pentagon ABCDE (where from now on we shall refer to the angle at a vertex by its letter and call the line-segment AB "b", BC "c" etc.), then we define "classes" of pentagons via relations between the angles or sides which constrain the kinds of pentagons one could form, e.g. "all pentagons with $a = b$, $d = e$, $A = 60°$ and $D = 120°$". In fact, this exact class along with another out of the fifteen distinct classes that exist are shown in Figure 1.4.

Michael Rao in 2017 announced that he had performed a clever series of deductions about the ways such pentagons could meet at a point to narrow down the possible valid classes to 371 and then, in a groundbreaking paper **Exhaustive search of convex pentagons which tile the plane**, he checked these cases with a computer program which showed that the 15 classes that had already been discovered were the only ones possible[5].

Thus, we have a complete classification of all convex prototiles that admit a

---

[3]Try and prove these facts - it's not that difficult and a bit of fun!

[4]This problem was in fact open when I first wrote this article!

[5]The tilings on the front cover of this chapter show one representative example from each class. The diagram was adapted from Rao's original paper.

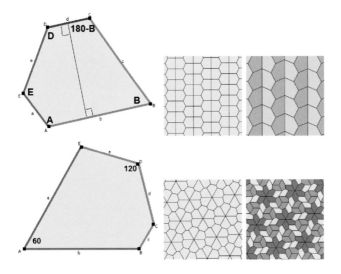

Figure 1.6: *Left*, two examples from the classes of pentagons drawn with colours indicating common lengths and angles labeled to illustrate relations, *right*, two examples of tilings involving pentagons from the corresponding class.

tiling of the plane. This naturally allows us to conclude a bunch of interesting simple facts, which mathematicians often call *corollaries*: firstly, we have an algorithm which solves the tiling problem for convex single-prototile tilings (check which number of sides it has; if it's bigger than 7 return "No", if it's smaller than 7 check if it fits in any of our known classes, if not return "No") and secondly, it showed that there is no single convex prototile that acts as an *aperiodic set*[6].

We've heard about aperiodicity quite a bit in this chapter and that's because for many years, hardly anything was known about them (Berger's initial discovery of a Wang tiling that was aperiodic involved 20426 prototiles!) until an important result was proved in 1998, which we shall endeavour not only to explain but to prove, by CHAIM GOODMAN-STRAUSS which led to a veritable zoo of examples and results.

Some tilings have one or more *substitution rules* - diagrammatic ways of dissecting tiles into collections of smaller tiles from the prototile set - so that by "chopping up" a tiling with large tiles, one obtains another valid tiling of the plane with smaller tiles. Given that it's possible to use the rules to make large tiles into small tiles using these rules, one might ask naturally whether

---

[6]Such a prototile is called an *einstein* and as of the time I'm writing this, nobody knows if one exists.

it's possible to go the other way with the same rules, i.e. group a collection of small tiles into a single larger tile using the substitution rules backwards (we'll call this process *compiling*).

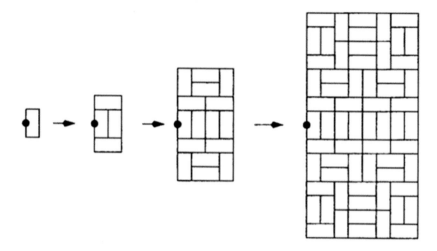

Figure 1.7: A diagram that illustrates an example of a substitution rule, taken from an influential 1998 paper on the subject.

This *is*, of course, often possible but, subtly, while dissection via the rules is always unique (you simply dissect each tile according to the relevant rule), compiling is not always unique (try and see if you can give an example of this yourself with just a square grid!). The celebrated theorem that Goodman-Strauss proved in his 1998 paper, **Matching rules and substitution tilings**, was that if a tiling with a substitution rule can be shown to have unique compiling also, then a tiling of the plane using its prototiles via substitutions will be aperiodic[7]!

Why? Let's suppose, hoping to reach a contradiction, that we had a tiling with unique compiling that was periodic. If you recall, periodicity definitionally means that there is some distance d and direction I can move the whole tiling such that it looks the same afterwards. Note that we can continue compiling the tiling forever (i.e. we compile it once to get a new tiling with larger tiles and then we can compile *that* tiling and then again etc.) and that each time the supertiles involved get larger - let's call our original tiling T_0 and the unique compiling of that T_1, the unique compiling of *that* T_2 etc. so that after n compilings we have T_n. Thus, if we look at some tile in T_0 and move it by our distance d, there is some number $m > 1$ of compilings we can perform to the tiling such that where

<hr />

[7]This statement is slightly simplified for the purposes of this article, as there is some subtlety as to what I mean by "via substitutions", but gets across the right idea. For those who are interested in such subtleties and have some time on their hands, I suggest looking up "matching rules" or simply reading Goodman-Strauss' paper.

the tile initially was *and* where it is after the movement both fall within the perimeter of a single supertile in T_m.

Now consider that because moving the patch by d left the tiling looking the same, T_m-moved-by-d must also be some iterated compiling of T_0. But because our tile was in the same supertile in T_m before and after the movement by d, this means that T_m and T_m-moved-by-d are distinct compilings of T_0, which violates our assumption that the tiling always has a unique compiling - contradiction!

So, if we assume that we have a periodic tiling with unique compiling we get a contradiction, so tilings with unique compiling cannot be periodic i.e. they are aperiodic.

That's quite a complicated argument, so I implore you read it through a few times or come back to it later, as it is extremely rewarding to understand *why* such an important principle is true. However, having done so, how does it help us? If we can find *any* prototile-sets with substitution rules that are unique both for dissection and compiling, then we know instantly that it's aperiodic[8], numerous such examples are shown for your entertainment in Figure 1.8!

If you've found this article interesting (or you just enjoy gazing at beautiful tilings), we suggest these topics for further reading:

1. ROGER PENROSE's contribution to aperiodic tilings & its connection to quasicrystals.

2. Wikipedia's wonderful catalog of Euclidean tilings by convex regular polygons

3. **The trouble with five** by Cambridge's *Plus Magazine*.

4. Voronoi tesselations

In addition, all of the aperiodic tilings mentioned in this article along with hundreds of more can be found at the remarkable *Tilings Encyclopedia*, which I highly suggest you look up! **Challenge I:**

1. The five-fold symmetry problem is to find a tile set that consists of polygons with exactly five-fold symmetry or to show that such a set cannot exist. Your goal is much simpler, though themed on that basis: Find one or more irregular pentagons that tile.

2. Find a prototile/shape of Heesch number 3. This is much harder, and I don't have the answer to this, so this is an effectively open problem!

---

[8]This is actually harder than it first sounds but is simple enough that mathematicians have since found hundreds of examples of aperiodic sets

Figure 1.8: Some great examples of aperiodic tilings, along with their substitution rules.

Chapter 2

Cellular Automata

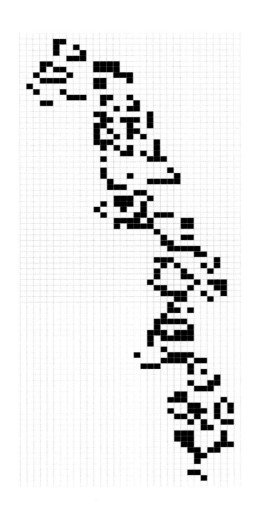

Salutations!

Welcome to the second instalment of ARM, this time taking a look at cellular automata. We have some exciting articles lined up for the future, potentially including but not limited to Chaos Theory, Trans-finite Set Theory, Game Theory and more - so, watch this space!

To understand what a cellular automaton is, I think it is helpful to understand the motivation behind the concept's creation in the 1940s, leading up to a book which essentially founded the field: **The Theory of Self-Reproducing Automata** by JOHN VON NEUMANN.

In this case, it is also necessary to have a word on John von Neumann, since he was an exceptional man and potentially merits the term genius in all the ways he contributed to mathematics. In fact, his remarkable intelligence was noticeable from 6 years old, when he was able to proficiently converse in Ancient Greek and became famous in his school for being able to multiply and divide 8-digit numbers in his head in under 10 seconds. By 15, he was under the tutelage of GABOR SZEGO (a mathematician famous for his contributions to calculus and linear algebra), had a great understanding of most founding fields of higher mathematics and at 19 years old, he published his first mathematical papers. Hans Bethe, who won the 1967 Nobel Prize for Physics for his contributions to our understanding of the nuclear reactions in the Sun, was a long-time friend of von Neumann's. When Bethe observed von Neumann talking with his baby daughter Marina, he noticed that "John" could very easily lower himself so that they were able to talk as equals and then wondered if von Neumann used just the same approach when talking to his physicist friends[1].

By the 1940s, von Neumann had become interested in self-replication. He wanted to understand on an abstract level how a dynamic system or mechanistic object like life on earth could go about creating copies of itself in an efficient way that could allow for variation (a property referred to as *evolvability*) – he came to the conclusion that it ultimately relied upon:

1. The propagation of information in systems.

2. The culmination of well-defined local rules into complex global behaviour.

3. The information structure (or "tape" as he called it) being an active part of the construction process.

Consider that these observations were made prior to the discovery of the structure of DNA - von Neumann had, in a heuristic sense, understood what it had to look like.

In order to understand in more depth the above ideas, he defined cellular automata as models, which would act as small and simple computing machines: consider an n-dimensional space (a plane, say) divided into discrete units (in our example of the plane, this would be a series of squares) where each unit is called a *cell*, which can be in one of a defined number of states. Then, we can

---

[1]I think this story is particularly remarkable if you stop to consider that nearly all of "his physicist friends" then were Nobel Prize winners or to-be winners

propagate each cell to a new state with the use of a *ruleset*, where for each cell we look at a collection of the cells surrounding it (called a *neighbourhood*) and based on their states, we assign that cell its "next" state. This sounds quite abstract, so to illustrate we can take a look at one of the so-called *elementary automata* as documented and investigated in the 1980s by STEPHEN WOLFRAM, shown in Figure 2.1.

Figure 2.1: A famous 1D automaton.

Figure 2.1 requires some explanation. Elementary cellular automata have 2 possible states, represented in the image by cells being either *black* or *white*, and are 1-dimensional, meaning that the neighbourhood their ruleset examines consists of a series of cells all in a line (in this case, the neighbourhood is the cell itself and the two squares adjacent to it). Each of the boxes below this automaton's name, "Rule 110", represent a rule from the ruleset: the row of three in a given box denotes a particular neighbourhood configuration and the cell in the middle column below it shows the defined outcome for that situation. For example, the far right box tells us that for this automaton, if a given cell is in the "white" state and the cells either side of it are also in the "white" state, then in the next iteration it will remain in the "white" state.

The image below the boxes shows what happens when you "run" the automaton: the first row, labeled "1" in the Figure, consists of just one cell in the "black" state and the rest in the "white" state; the row labeled "2" shows the outcome of applying the automaton's ruleset to row 1 and row "3" shows the output of row "2" etc. If we were to consider each application of the rules to be a step forwards in time, then Figure 2.1 essentially has "space" on the x-axis and "time" on the y-axis going downwards.

You might have noticed by looking at the rule boxes carefully that if you imagine the "black" state to represent a "1" and the white state to represent a "0", then the situations being considered in each rule count down from 111 in binary from left to right. The idea behind this systematic referencing of situations was to create a general notation system for cellular automata: for an n state cellular automaton, arrange all the possible situations in the neighbourhood of a given cell in boxes like those in this image (except in this case, you would be counting down in base-n rather than just base-2) and assign them each a number from 1 to n as an "outcome state". Then, by writing all of these outcome state numbers next to each other in the order of the boxes on a piece of paper, we have a number written in base-n which uniquely identifies this automaton. This sounds initially complex but it's quite intuitive when you wrap your head around it. For this automaton, simply do what I said: the rule boxes are already written out for us and the digits below each box are "0" if the outcome state is "white" and "1" if the outcome state is "black", so when we write each of these digits out in the order of the boxes we get "01101110", which in base-10 is the number 110 (hence the name, *Rule 110*). These numbers are called *Wolfram numbers* after the aforementioned Wolfram began to catalogue thousands of cellular automata to understand them experimentally using this notation.

Why did I mention this cellular automaton? It just seems to be a bunch of random rules I chose and then simulated... In fact, it is computationally universal or, more formally, *Turing-complete*. What does this mean? Matthew Cook, in the 2004 paper **Universality in Elementary Automata**, found that given *any possible* computer program and a set of inputs for it, he could encode (with some quite complex and specific constructions) this data into the states of a 1D cell-tape such that by running Rule 110 on it, he would simulate the computations of the program - so essentially, Rule 110 can simulate *any* computational process! While the proof itself and perhaps even the proper statement of the theorem is quite complicated and beyond the scope of ARM, we can show you illustrations of information transfer within Rule 110 to give you an idea of how such an encoding might work.

Cook identified and categorised structures that moved periodically, which he called "spaceships", in a diagram taken from his paper in Figure 2.2.

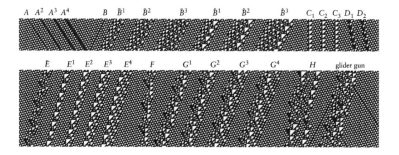

Figure 2.2: All Rule 110's spaceships and their names according to Cook.

Figure 2.2 consists of a similar diagram to that of Figure 2.1 with time going downwards on the y-axis and space on the x-axis. The labels above each structure are the names that Cook assigned them in his paper.

He also examined in some detail how they collide, some of which are shown in Figure 2.3.

Precisely who out of Cook and Wolfram are responsible for this proof is a matter of continuing controversy which the reader may be interested in researching.

We must return to John von Neumann and his design for a self-replicating

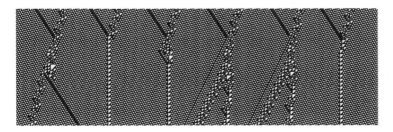

Figure 2.3: The six possible collisions between an A^4 and an $\bar{E}$

machine: while Rule 110 is only 1 dimensional with 2 states, von Neumann's original automaton had a grand *29* states, was *2* dimensional & used a neighbourhood which looked at a cell and its four adjacent cells. As you can imagine, the ruleset would be massive and is too long to explain here (it required its own book) but we can give an outline. There was the *null or ground* State 0 as per usual, a group of 8 *transmission states* which could act as wires to transmit binary signals, a group of 4 *confluent states* which could act as junctions for these wires to give information to & finally 16 *transition states* which could be built by confluent cells given the right signals from transmission cells.

The result was an incredible piece of engineering (one which many have tried fairly unsuccessfully to make simpler): a universal constructor, which, if added to any arrangement of cells in von Neumann's automaton, would be able to reproduce that arrangement countless numbers of times. The image in Figure 2.4 was constructed many years after von Neumann's death due to the work of a mathematician who actually implemented von Neumann's machine. Considering the fact that von Neumann made this without the existence of proper computers that could run simulations of cellular automata, this is undoubtedly impressive.

To explain Figure 2.4: the left part of the image shows the constructor in some detail, with each of the colour representing various of the complex zoo of states we described earlier; the right part of the image shows a series of constructors where the lowest one labeled "Generation 1" created the constructor labeled "Generation 2" and so on. Generation 3 had an added drawing of a flower put near it to demonstrate that, it being a replicator, in the Generation 4 constructor this "mutation" is passed on and the flower is replicated along with the main structure.

Many mathematicians have considered this work quite astounding due to the existence of the so-called *Garden of Eden patterns*, which are a set of cells with specific states which cannot be formed through iteration of a previous set (though they can clearly still act as a starting "position"). He apparently accounted for this in his design and since then a much greater understanding of Gardens of Eden has come about due to a theorem, proved by the mathematicians EDWARD MOORE & JOHN MYHILL, which states that such patterns exist for a given cellular automaton if and only it is possible using cells of the automaton to

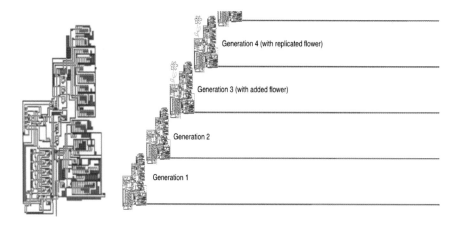

Generation 4 (with replicated flower)

Generation 3 (with added flower)

Generation 2

Generation 1

Figure 2.4: von Neumann's *universal constructor* along with an added drawing of a flower showing that it too is replicated in each subsequent generation.

construct pairs of structures which after an iteration become the same structure.

Due to the relative obscurity of the work and the fact that his book was put in the shadow by his many other great works, von Neumann's universal constructor and indeed the whole idea of cellular automata was forgotten for around 30 years until a British mathematician JOHN CONWAY (a recurring character in this book) created a much simpler 2-state, 2-dimensional automaton whose neighbourhood is the 3×3 block of cells around-and-containing a given cell and which exhibited surprising complexity. This automaton was called *Conway's Game of Life* and its rules were the following:

- A State-1 (alive) cell with fewer than 2 State-1 cells in its neighbourhood becomes State 0 ("dies").

- A State-1 cell with 2 or 3 State 1 cells in its neighbourhood stays at State-1.

- A State-1 cell with more than 3 State-1 neighbours becomes State-0.

- A State-0 cell with exactly 3 State-1 neighbours becomes State 1.

With just these four rules come the surprising results that the automaton:

- Is capable of universal construction, despite being so much simpler than von Neumann's automaton,

- Is capable of universal computation, like Rule 110,

- Has the property that, in general, the problem of determining whether the system eventually will evolve to a "static" situation (where no further changes happen to the states of any cell in future iterations of the system) is *generally undecidable*, i.e. there is no algorithm that can solve this problem for all situations.

As Conway said himself, "My life game wasn't *designed*. I just sort of thought that if you couldn't predict what it did, then probably that was because it was capable of doing anything." There are many complicated structures, such as analogues of the previously mentioned spaceships, in the Game of Life. For example, in Figure 2.5 we see the sequence of iterations of the famous "Glider".

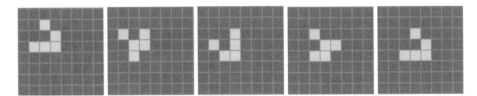

Figure 2.5: The Glider moving.

This object will move down diagonally right – I suggest you take out a piece of square paper and check with the rules I mentioned! Another option is to look it up on *Life Wiki*, the site made by enthusiasts of the automaton which catalogues thousands of structures made in C's GoL and for each one, has an interactive simulator so that you can watch it "go".

Even more amazing, people have made spaceships which release objects that release gliders (as in, they make glider guns) themselves! Images of these sorts of contraptions can be seen in Figure 2.6 at the end of the article.

Finally, we should consider the work of CHRISTOPHER LANGTON, a somewhat famous mathematician and computer scientist who founded the field of *Artificial Life* and made significant contributions to the field of cellular automata (for example, creating the Turing-complete *Langton's Ant*).

He once described an automaton *invariant* (an invariant is a kind of constant statistic that tells you something about an object) which he named his *lambda parameter*. One calculates this simply by taking the ratio r of the number of configurations of the neighbourhood that lead to a "live" or nonzero state to the number of possible configurations of the neighbourhood and, while simple to calculate, it brings many revelations.

What Langton observed was that for values of λ near 0, a randomly chosen automaton would often be "ordered" – i.e. they would either die out very quickly or they settle into predictable patterns - and for values near 1, the automaton would generally display highly chaotic behaviour. He finally observed that the majority of the complex automata, which would show groups of cells acting as discrete structures and interacting with each other, would often have a lambda value close to 0.5, the boundary which he referred to as the "edge of chaos".

Some people have suggested that this is analogous to the brain's function and life itself: not quite chaos, definitely not simply ordered but somewhere inbetween. Others have suggested that the process of natural selection actually pushes chemical reactions (which, when considered abstractly are very similar to cellular automata) to the edge of chaos, since the majority of chemical reactions happen ambiently, when certain imbalances in charge or fluids occur in a variety of media, but those whose mechanics are slightly more complex & change the environment around them in such a way as to continue propagating are by definition more likely to "survive" – potentially suggesting that biogenesis is both natural and inevitable. Others, however, discredit these ideas as unmathematical, saying that their lack of concreteness is a sign of their lack of meaning. We leave it to the reader to decide which viewpoint is correct.

If you've found this edition of Adventures in Recreational Mathematics and/or want to play around with other cellular automata, we suggest these topics for further reading:

- Finite-state Machines (part of understanding the more formal definition of C.A.)

- Von Neumann Universal Constructor

- Wireworld: another, extremely intuitive C.A. in which some individuals have made a fully programmable computer!

- Langton's Loops, Langton's Ant & Turmites in general. . .

- A New Kind of Science by Stephen Wolfram and reviews by various scientists or mathematicians.

- Garden of Eden Theorem

- Rule 30

Challenge II:

1. Calculate the λ-value for the Game of Life.

2. Design your own spaceship in your own cellular automaton. There are quite literally an uncountably infinite number of ways of answering this one, so give it some thought.

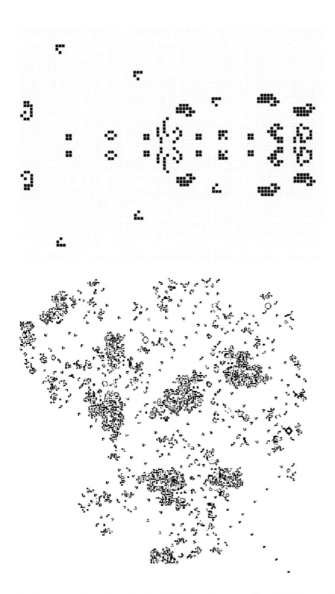

Figure 2.6: *Left*, a complex "puffer" which emits a trail of gliders either side as it moves, *right*, an extremely complicated "glider gun".

Chapter 3

Confessions of a Sequence Addict

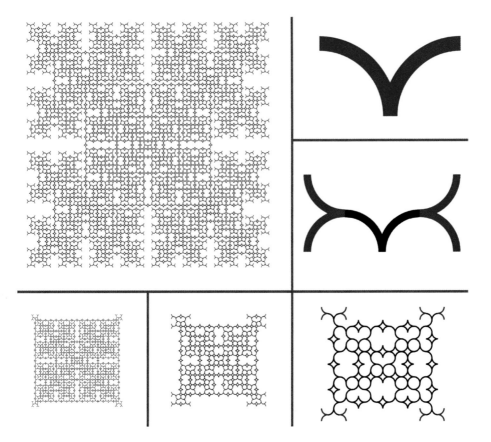

Salutations!

Welcome to the third instalment of Adventures in Recreational Mathematics! This time we will be discussing 4 *integer sequences* which have very interesting properties - it's an intriguing puzzle to see the first few terms of a sequence of numbers and to try and guess the rule behind it... For those in the earlier years of their education, they will encounter such sequences as 1, 3, 5, 7, 9, 11 ... *the odd numbers* or 1, 2, 4, 8, 16, 32 ... *powers of two* but in this issue we will look at rules that are far more complex and lead us on brilliant conceptual journeys, full of surprises.

We begin with a self-referential sequence described by WILLIAM KOLAKOSKI in 1965. Its first few terms are

$$1, 2, 2, 1, 1, 2, 1, 2, 2, 1, 2, 2, 1, 1, 2, 1, 1, 2, 2, 1, 2, 1, 1, 2, 1, 2, 2, 1, 1, ...$$

Can you guess how it works? In order to illustrate the rule for generating this sequence, imagine that we wish to construct a new sequence from the terms of this *Kolakoski sequence* by writing down the lengths of the "runs" of consecutive integers in the sequence which are the same: we first write down "1", because the first term of the K. sequence is 1 and the second term is already different, then we write down "2", because there are two 2s next to each other before we reach the next "1", then we write down "2" again, because there are two 1s next to each other etc.

What you will find, if you continue this, is that this process generates the same sequence!

It is this property that determines the Kolakoski sequence - it is the unique integer sequence using integers from the ordered list 1, 2 (in the order that they appear in the list) such that it is *its own run-length encoding.*

There are many questions that arise when discovering something like this: given only the beginning segment above of the sequence and being told the rule, could we have deduced naturally that this sequence only contain 1s and 2s? Yes: it can be seen from the fact that new terms of the sequence are generated from the runs of previous parts of the sequence and since no runs of a length greater than two appear in the segment above, we can deduce that no numbers higher than 1 or 2 will ever appear. Secondly, is this sequence unique in this recursive property or if we allowed other sets of integers could we generate more? One observation you might make when thinking about this question is that the sequence still satisfies the "run-length" requirement if you omit the first term so there are at least *2* such sequences, but we can do better than that.

There is in fact an infinite family of these sequences which use different integer lists. For example:

- $1, 2, 3$ generates 1, 2, 2, 3, 3, 1, 1, 1, 2, 2, 2, 3, 1, 2, 3, 3, 1, 1, 2, 2, 3, 3, 3, 1, 2, . . .

- $1, 3$ generates 1, 3, 3, 3, 1, 1, 1, 3, 3, 3, 1, 3, 1, 3, 3, 3, 1, 1, 1, 3, 3, 3, 1, 3, 3, . . .

We can also use *infinite* lists: consider using the primes $2, 3, 5, 7, 11, ...$, which generate 2, 2, 3, 3, 5, 5, 5, 7, 7, 7, 11, 11, 11, 11, 11, 13, 13, 13, 13, 13, 17, 17, 17, 17, 17, ... or just using the integers *themselves* - $1, 2, 3, 4, 5, ...$ generate the sequence 1, 2, 2, 3, 3, 4, 4, 4, 5, 5, 5, 6, 6, 6, 6, 7, 7, 7, 7, 8, 8, 8, 8, 9, 9, ... which also has the property that the nth term is the number of times that n appears in the sequence.

Something you can check is that by using different orders of the integers in your sets, you can create different Kolakoski sequences - try generating one with pen and paper using the integers $3, 1, 2$ *in that order* and contrast this with the sequence generated by $1, 2, 3$ above. You may be able to believe, by now, that *any* list of integers in which no two consecutive terms are the same (which, if the list is finite, includes the *first* and *last* elements not being the same) can generate its own Kolakoski sequence. Quite remarkable.

If we call the process of writing down the *run-lengths* of a sequence as above "*Kolakosking* a sequence" then we can ask another natural question: we've found a bunch of sequences that are their own Kolakoskings, but are there sequences with the property that by Kolakosking them twice (and no fewer than twice) you get the same sequence?

Once again, it is amazingly possible to find examples of this: a sequence S_1 whose run-lengths generate a new sequence S_2 whose run-lengths generate S_1! Here are two examples: A pair which generate each other:

- 1, 1, 2, 1, 1, 2, 2, 1, 2, 2, 1, 2, 1, 1, 2, 2, 1, 2, 2, 1, 1, 2, 1, 2, 2, ...

- 2, 1, 2, 2, 1, 2, 1, 1, 2, 2, 1, 2, 2, 1, 1, 2, 1, 1, 2, 1, 2, 2, 1, 1, 2, ...

and a miraculous cyclical quintuplet where each sequence in the list generates the one below it:

- 1, 1, 2, 2, 3, 3, 4, 4, 4, 5, 5, 5, 1, 1, 1, 2, 2, 2, 2, 3, 3, 3, 3, 4, 4, ...

- 2, 2, 2, 3, 3, 3, 4, 4, 4, 5, 5, 5, 1, 1, 1, 1, 2, 2, 2, 2, 3, 3, 3, 3, 4, ...

- 3, 3, 3, 3, 4, 4, 4, 4, 5, 5, 5, 5, 1, 1, 1, 1, 2, 2, 2, 2, 3, 3, 3, 3, 3, ...

- 4, 4, 4, 4, 4, 5, 1, 1, 2, 2, 3, 3, 3, 4, 4, 4, 5, 5, 5, 5, 1, 1, 1, 1, 2, ...

- 5, 1, 2, 2, 3, 3, 4, 4, 4, 5, 5, 5, 1, 1, 1, 1, 2, 2, 2, 2, 3, 3, 3, 3, 4, ...

Don't just take my word that they work, try them! Can you, using this quintuplet as an example, figure out a way of generating an n-tuplet where each sequence uses all the integers from 1 to n?

A final question you may find yourself asking is: does the Kolakoski sequence ever repeat? Nobody knows...

Self-description can be encoded mathematically in many ways, often leading to complex or chaotic behaviour - Kolakoski's sequence used and *Gijswijt's sequence*, named after the mathematician DION GIJSWIJT, uses another.

To explain the process of determining the $(n + 1)$th term of the sequence from the first n terms, one needs to understand strings and basic string notation. A *string* is simply a series of one or more characters (not limited to numbers) written next to each other just like letters in a word. The process of taking a collection of strings and writing them one after the other into a single string, such as taking the strings "b0at" and "h0u$£" to make "b0ath0u$£" (the use of alternative characters here is to emphasise how they are not limited to numbers or letters), is referred to as *concatenation*. If Y is a string, then we denote Y concatenated with itself k times as Y^k.

Thus, given the first n terms of Gijswijt's sequence we can concatenate them into a string and identify sections of it X, Y such that the whole string is equivalent to XY^k for some integer k. Suppose that we then choose Y such that it is the *longest possible string* to satisfy the previous conditions, then we define $(n+1)$th term of the sequence to be the corresponding k. To illustrate, let us go through the creation of the first few terms together.

Originally, we have nothing and since that nothing is "repeated once" (it is the Y for this string), we write "1" as the first term. Since 1 is repeated only once so far, we write another "1". Now, as there are two 1s in succession, we write down "2". See if you can understand the rest here:

$$1, 1, 2, 1, 1, 2, 2, 2, 3, 1, 1, 2, 1, 1, 2, 2, 2, 3, 2, 1, 1, 2, 1, 1, 2, ...$$

Remember that we are choosing Y to be longest - in the situation where we only had the terms $1, 1, 2, 1, 1, 2$ we could have chosen $X = $ '11211' and $Y = $ '2', in which case $k = 1$, but by choosing $X = $ '' and $Y = $ '112' we get a longer Y and so $k = 2$ (which is what we see written in the above sequence).

Reasoning about it isn't simple at first – for example, notice how after the 2 beyond the two 1,1,2 blocks, another "2" appears because of the two 2s next to each other and then, unexpectedly, a 3 appears because of the three twos in a row... In fact, it is this nontrivial "block and glue" structure that is key to its understanding:

Let's refer to Gijswijt's Sequence as G_1 and then observe that the entire sequence is made up of "blocks" that are doubled and then are ended with some sort of "glue" sequence, for example the first block B_1 is 1 and the first glue E_1 is 2 since we begin with "1,1,2". The next block, B_2, is 1,1,2 (or $B_1B_1E_1$) and the next glue sequence E_2 is 2,2,3. We can generalise this to larger blocks and glue sequences to say that $B_{n+1} = B_nB_nE_n$ where E_n is simply defined to be the sequence of numbers starting at the end of B_nB_n and ending at the first 1 reached.

What's interesting about this analysis is that one can combine all the glue sequences into a new infinite sequence, G_2, which follows the same $(n + 1)$th term rule as G_1 *except* instead of writing the k from XY^k, we write whichever the larger number is out of the set $\{2, k\}$. Here are the first 25 terms:

$$2, 2, 2, 3, 2, 2, 2, 3, 2, 2, 2, 3, 3, 2, 2, 2, 3, 2, 2, 2, 3, 2, 2, 2, 3, \ldots$$

One can perform similar structural analysis on the "block and glue" structure of G_2 as above and further construct a G_3 and more generally, G_n. We can observe that every G_n starts with n and so, perhaps surprisingly, see that Gijswijt's Sequence is unbounded (that is, there is no finite positive integer that does not appear at some point in the sequence). However, it should be noted that the sequence grows exceptionally slowly, with n believed to generally appear at roughly the $2^{3^{4^{5^{\cdots^n}}}}$ th term and thus, despite its unboundedness, is conjectured to have a *finite average*!

————— ⟋⟍ —————

Our next phantasmagorical sequence was first investigated by the lesser known (and yet important) combinatoricist and number theorist JAMES JOSEPH SYLVESTER[1]. It begins

$$2, 3, 7, 43, 1807, 3263443, 10650056950807, 113423713055421844361000443, \ldots$$

Given that the sequence shows such incredible growth, can you guess what the rule for making the next term is?

One constructs the $(n+1)^{th}$ term by taking the product of the first n terms and adding one – you can see this above. Equivalently, one can define the nth term of the sequence (s_n) in terms of the preceding term as $s_n = s_{n-1} \times (s_{n-1} - 1) + 1$ with $s_0 = 2$. Can you see why these definitions are equivalent?

Let us consider the sum of the reciprocals of the terms in what we shall call *Sylvester's sequence*. We can first see definitionally that

$$\frac{1}{s_i - 1} - \frac{1}{s_{i+1} - 1} = \frac{1}{s_i - 1} - \frac{1}{s_i(s_i - 1)} = \frac{(s_i - 1)}{s_i(s_i - 1)} = \frac{1}{s_i}$$

Thus, we can say

$$\sum_{i=0}^{j-1} \frac{1}{s_i} = \sum_{i=0}^{j-1} \left(\frac{1}{s_i - 1} - \frac{1}{s_{i+1} - 1} \right)$$

$$= \frac{1}{s_0 - 1} + \left(-\frac{1}{s_1 - 1} + \frac{1}{s_1 - 1} \right) + \left(-\frac{1}{s_2 - 1} + \frac{1}{s_2 - 1} \right) + \ldots - \frac{1}{s_j - 1}$$

$$= \left(\frac{1}{s_0 - 1} - \frac{1}{s_j - 1} \right) = \left(\frac{1}{2 - 1} - \frac{1}{s_j - 1} \right) = 1 - \frac{1}{s_j - 1} = \frac{s_j - 2}{s_j - 1}$$

We can deduce from the last two equivalences an impressive result: since the nth terms of Sylvester's sequence tend to infinity as $n \to \infty$, the sum of

---

[1] Who, incidentally, is the individual who introduced the word "matrix" into mathematics.

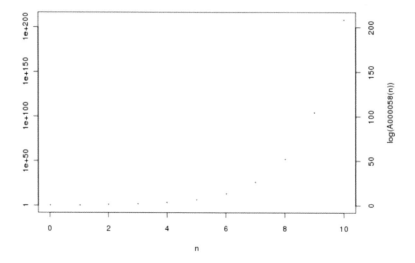

Figure 3.1: A graph of the log of the first few terms of *A000058*, the OEIS's (explained later) identifier for Sylvester's sequence. Note the non-linearity of the graph even with the logarithmic scale, displaying that the sequence grows *doubly exponentially*.

the reciprocals of its terms tends to 1! Furthermore, the last equivalence in particular tells us that if we were to sum up to the jth term in the sequence and subtract one from the last denominator, we would obtain a value of 1 also, since $\frac{s_j - 2}{(s_j - 1) - 1} = 1$.

Thus, we can use the sequence to create an infinite series of representations of 1 as a sum of fractions whose numerator is 1:

$$1 = \frac{1}{2} + \frac{1}{3} + \frac{1}{6}$$

$$1 = \frac{1}{2} + \frac{1}{3} + \frac{1}{7} + \frac{1}{42}$$

$$1 = \frac{1}{2} + \frac{1}{3} + \frac{1}{7} + \frac{1}{43} + \frac{1}{1806}$$

$$\cdots$$

This allows us to have a different interpretation of Sylvester's sequence: a "greedy algorithm" which at each stage chooses the smallest n such that $\frac{1}{n}$ plus the sum of the reciprocals of all the terms so far is less than one. Thus, remarkably, Sylvester's sequence has the property that its infinite sum of reciprocals is the fastest-converging series of unit fractions to 1! !

The new interpretation also allows us to consider other "Sylvestrine" sequences. For example, what about the greedy algorithm series formed from $s_0 = \frac{1}{4}$ and the limit being 2? I was able to compute these 6 terms using a quickly made computer program:

$$2, 2, 5, 21, 421, 176821, \ldots$$

Do you notice anything? The rule that generates terms of Sylvester's sequence also applies here!

If one tries to derive similar convergent sequences to rational numbers using the greedy algorithm, one will often observe a Sylvestrine relation; this can all be explained by a result obtained by a modern mathematician, PAUL ERDOS, which states that if one has a sequence of integers a_n such that $a_n \geq a_{n-1}^2 - a_{n-1} + 1$ and the sum of its reciprocals converges to some rational number then after some point in the sequence, the terms will follow the relation that defines Sylvester's sequence. So, Sylvester's sequence is not only interesting as specific example but, in fact, there is something fundamental about its defining rule to doubly-exponential sequences...

Investigations into its closed-form representation (i.e. one that does not make reference to previous terms to calculate the nth term) have led to the incredible discovery that there exists a real number E, the *Sylvester Constant*, such that

$$s_n = \left\lceil E^{2^{n+1}} + \frac{1}{2} \right\rceil$$

where $\lceil x \rceil$ denotes the number you get when you *round up* x to the nearest integer and

$$E = \frac{\sqrt{6}}{2} \exp\left\{ \sum_{j=1}^{\infty} 2^{-j-1} \ln\left(1 + (2s_{j-1} - 1)^{-2}\right) \right\} \approx 1.26408$$

It's also of interest that although the sum of its reciprocals is 1, if we take the alternating sum of its reciprocals (i.e. $\frac{1}{s_0} - \frac{1}{s_1} + \frac{1}{s_2} - \ldots$) then the result is a *transcendental number*. This means that not only is it an *irrational* number, it is not the root of *any polynomial*[2].

We end off with a fun sequence investigated by the British mathematical *jack-of-all-trades*, JOHN HORTON CONWAY. He recalls that a student once excitedly brought a sheet of paper in front of him with these 5 terms on it:

$$1, 11, 21, 1211, 111221, \ldots$$

and he was asked to guess the next number in the sequence. He wasn't able to do this - can you?

---

[2]For those less familiar with the terminology, a *polynomial* is an algebraic expression of the form $a_0 + a_1 x + a_2 x^2 + \ldots a_n x^n$ where $a_0, a_1, \ldots, a_n$ are integers and a *root* of the polynomial is a number R such that $a_0 + a_1 \times R + a_2 \times R^2 + \ldots + a_n \times R^n = 0$.

After some guesses and deliberation the student revealed to him the rule for constructing the rest of the sequence – the nth term of what he called the *Look-and-Say numbers* was made by reading through the $(n-1)$th term and counting those groups of consecutive numbers which are the same. For example, starting with "1" we would say that there is "One 1" and thus the next term is "11". For something more complicated like "1211", we say there is one "1", followed by one "2" and then a final grouping of two "1"s: thus "11 12 21" concatenated, or "111221".

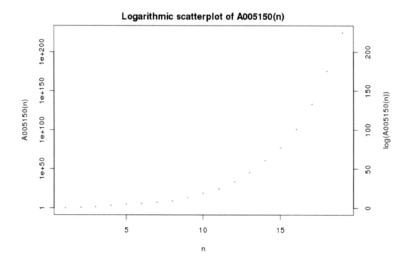

Figure 3.2: A graph of the first few terms of *A005150*, the OEIS identifier for the Look-and-say sequence.

Conway took some time later to investigate the properties of this seemingly arbitrary recurrence relation and the many sequences that one could generate from different starting integers. What was immediately apparent was that any sequence apart from the anomalous $22, 22, 22, ...$ would tend to infinity and that for any starting digit $n \neq 1$, the sequence would be

$$n, 1n, 111n, 311n, 13211n, ...$$

with n forever staying at the end. He then came upon and proved two remarkable results, which he published in his brilliantly titled paper **The Weird and Wonderful Chemistry of Audioactive Decay**. The first was the celebrated *Cosmological Theorem* which states that after some finite point, any look-and-say sequence splits into a discrete grouping of strings whose dynamics are *independent* beyond this point (i.e. continue to evolve just as they would if the rest of the numbers weren't there). It turns out that there are only 92 such strings involving the digits 1s, 2s & 3s, so he called them "elements" and referred to this process of sequence bifurcation as *audioactive decay*. The elements involving digits other than just 1, 2 or 3 were called "transuranic" since their abundance in a given n-length string tends to zero as n tends to infinity (they don't change beyond just appearing as "$1n$").

The table of elements is too large to be shown here (having 92 rows!) but I urge you to look it up. After all, the mechanics of any look-and-say sequence is beyond a point (maximally, the 24th term) completely determined by these elements and furthermore, by examining the lengths of all the elements, we can precisely predict the change in the length of the term.

Suppose we construct a 92×92 matrix A in which entry $A_{x,y}$ (the entry on the xth column and yth row) is equivalent to the product of the number of times the xth audioactive-element appears in the decay-successor to the yth audioactive-element and the ratio of their lengths. Then consider representing the length of a given term T in a look-and-say sequence as a vector whose nth coordinate-entry is the length of the nth audioactive-element multiplied by the number of times it appears in T.

By multiplying the "length vector" by the matrix, we get a new vector which encodes the length of what would be the next term in the look-and-say sequence (because the sum of its entries is what the length would be!) so this matrix is what mathematicians call the *transition matrix* of the lengths of terms in look-and-say sequences. By examining this matrix, he deduced a new theorem which seems utterly bizarre (before one learns about some of the ways of analysing matrices that are beyond the scope of this article).

If one looks at look-and-say sequence terms' lengths, you can see that they increase exponentially. Figure 2.3 is a logarithmically-scaled graph of the length of terms in the traditional look-and-say sequence starting with 1.

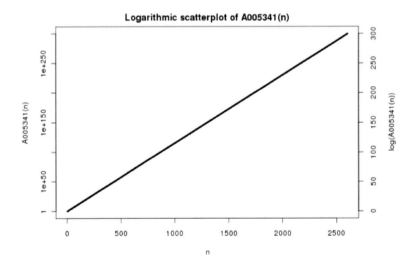

Figure 3.3: A graph of the logarithm of the first few terms of $A005341$, the OEIS identifier for the length of the nth term of the classic Look-and-say sequence.

The amazingly straight line means that the length of terms scales by an almost constant value across the sequence. This is intriguing – basic investigation reveals that it is a roughly 30% increase each term. Conway was able to show that, as $n \to \infty$, the ratio of the lengths of the $(n+1)$th and nth terms does in fact tend to a constant and found a way of determining its value. It's the unique real root of the polynomial of degree 71 displayed in Figure 3.4 and is roughly equivalent to 1.3036.

$$
\begin{aligned}
&x^{71} - x^{69} - 2\,x^{68} - x^{67} + 2\,x^{66} + 2\,x^{65} + x^{64} - x^{63} - x^{62} - x^{61} - x^{60} - x^{59} + 2\,x^{58} + \\
&5\,x^{57} + 3\,x^{56} - 2\,x^{55} - 10\,x^{54} - 3\,x^{53} - 2\,x^{52} + 6\,x^{51} + 6\,x^{50} + x^{49} + 9\,x^{48} - 3\,x^{47} - \\
&7\,x^{46} - 8\,x^{45} - 8\,x^{44} + 10\,x^{43} + 6\,x^{42} + 8\,x^{41} - 5\,x^{40} - 12\,x^{39} + 7\,x^{38} - 7\,x^{37} + 7\,x^{36} + \\
&x^{35} - 3\,x^{34} + 10\,x^{33} + x^{32} - 6\,x^{31} - 2\,x^{30} - 10\,x^{29} - 3\,x^{28} + 2\,x^{27} + 9\,x^{26} - 3\,x^{25} + \\
&14\,x^{24} - 8\,x^{23} - 7\,x^{21} + 9\,x^{20} + 3\,x^{19} - 4\,x^{18} - 10\,x^{17} - 7\,x^{16} + 12\,x^{15} + 7\,x^{14} + \\
&2\,x^{13} - 12\,x^{12} - 4\,x^{11} - 2\,x^{10} + 5\,x^{9} + x^{7} - 7\,x^{6} + 7\,x^{5} - 4\,x^{4} + 12\,x^{3} - 6\,x^{2} + 3\,x - 6.
\end{aligned}
$$

Figure 3.4: A large polynomial

This astounding result suggests that the look-and-say generation rule, far from being arbitrary, is apparently very ordered. It led to Conway proclaiming that this investigation was the "most complicated solution to the simplest problem" that he had ever seen - you can see in Figure 3.5 the various other roots of the polynomial mentioned.

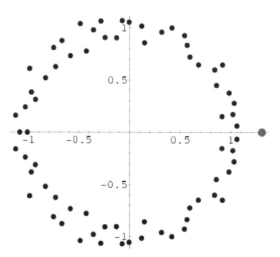

Figure 3.5: The roots of the *characteristic polynomial* of the matrix mentioned above plotted on the complex plane - the red point represents the value of the length-ratio constant.

To conclude, I wanted to mention an amazing resource which slips beneath the radars of full-time and recreational mathematicians alike at our age: the *Online Encyclopaedia of Integer Sequences*. Nearly all the terms and sequences mentioned in this article I only found out about due to the OEIS; most of the graphs in this article were taken from the OEIS's graph tool. The OEIS was set up in the 2000s as the result of a 40-year project that had been run by NEIL SLOANE, a number theorist particularly interested in integer sequences – indeed, the title of this article is taken from a talk of the same name given to a computer science group at Princeton University in 2016.

The OEIS has reference codes for all its sequences in the form "AXXXXXX", which allow you to easily find what you are looking for on their website. For further reading or your own investigations, here are the OEIS codes for the sequences mentioned in this article:

- Kolakoski Sequence, A000002

- Gijswijt's Sequence, A090822

- Sylvester's Sequence, A000058

- Look-and-say Sequence (starting with 1), A005150

If your appetite has been whetted for more interesting sequences, perhaps look up such beauties as the *Recaman sequence* and the *Kissing number sequence* or investigate what *A000001* is and why *it*, out of all those that could have been chosen, was given the special first entry in the database...

Finally, we couldn't finish an ARM article without challenges.

Challenge III:

1. Investigate other "Sylvestrine" greedy sequences as mentioned – can you create an infinite number of different representations of the number they converge to as finite sums of unit fractions like above?

2. Create a formula for the nth term of the traditional Kolakoski Sequence (it can be recursive!). Using this, record the relative frequency of 1s and 2s in the sequence up to some large n via a computer program implementation. Make a conjecture of what fraction of the sequence's terms are 1s based on this.

Chapter 4

Cantor's Attic

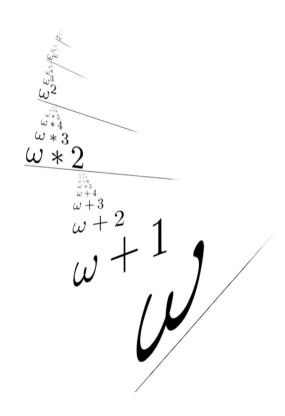

Salutations!

Welcome to the 4th installment of Adventures in Recreational Mathematics! This article is about GEORG CANTOR's work on the puzzling subject of infinity. Amongst other things, his work led him to believe some infinite sets were smaller than others, he believed there are an infinity of infinities and that his work and infinity came from God. His work and ideas were revolutionary and came under attack from such people as Wittgenstein, calling his work "utter nonsense", Poincaré, who said Cantor's ideas constituted "a disease infecting mathematics" and worst of all KRÖNECKER, who described him as "a corrupter of youth, a scientific charlatan". Despite these fierce attacks from Premier League detractors, today his work is generally accepted - read on to see what you make of his revolutionary and counter intuitive ideas.

We begin with the concept of an ordering. A set is a collection of objects; those objects can be "ordered" according to some relation - for example, if our set contains people we can sort them by size, by age, alphabetically etc. Formally this is stated as a binary relation R relating two objects in a set[1] (which is written xRy) so that for all elements a, b, c in the set it satisfies these three axioms:

1. aRa (reflexivity)

2. If aRb and bRa, then $a = b$. (antisymmetry)

3. If aRb and bRc, then aRc (transitivity)

R places the objects of a set into a hierarchy. R is called a *partial order* and the set it governs is naturally called a partially ordered set (*poset*). From now, we will use the notation $x \preccurlyeq y$ to say that x is lower than y in the hierarchy for some poset, as partial orders have enough restrictions to be drawn in a similar manner to the $\leq$ symbol (which is an example of a partial order over the positive integers).

To demonstrate how partial orders don't have to exhibit a linear hierarchy, we can visualise posets using *Hasse diagrams*, where we draw a node for every element of the poset and then draw a line *upwards* from an element x to an element y if $x \preccurlyeq y$. Below, in Figure 4.1, we can see two such diagrams: the leftmost displays the set F of factors of 120, where $x \preccurlyeq_F y$ if x is a factor of y, and the rightmost shows the set of all subsets of $\{2, 3, 4, 5, 8\}$ such that if x is a member of one such subset then all members of $\{2, 3, 4, 5, 8\}$ which are factors of x are also in the subset, ordered by *set inclusion*.

Do you notice anything interesting? Although the subject matters of the relations involved in the posets are quite different, their Hasse diagrams (and thus their orderings) appear to have, in some sense, the *same structure*. To be more precise, let's consider two posets and their corresponding order relations $(S, \preccurlyeq_S)$ and $(T, \preccurlyeq_T)$. A bijective function (a one-to-one mapping between the elements of two sets) f from S to T is called an *order isomorphism* if $x, y \in S, x \preccurlyeq_S y$ implies $f(x) \preccurlyeq_T f(y)$ and vice versa; the existence of such a function

---

[1]If this seems obscure, you know of many binary relations already such as $=, \leq, \subseteq$.

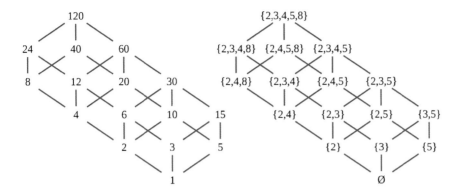

Figure 4.1: An example of order isomorphism.

means that S and T are order isomorphic. This is an important concept and based on the mass of symbols you just saw sounds extremely complicated - all it really says is that two of posets are order isomorphic if you can relabel one's Hasse diagram to look like the other's with the requirement that no two elements from the first are relabeled with the same name.

Well-orders paired with *well-ordered sets* (wosets) make up a subset of partial orders & posets which have a few more restrictions/axioms they must satisfy:

- For any two elements a, b of a woset $(W, \preccurlyeq_W)$, one of $a \preccurlyeq_W b$ and $b \preccurlyeq_W a$ is true.

- In any collection of items from our W (i.e. any subset), there must be an element c which is "below" (in the ordering-hierarchy of $\preccurlyeq_W$) any other element in the rest of the collection.

These properties of relations are called *totality* and *well-foundedness* respectively and the first of them in particular may strike you as odd. Why would we need to specify that for any two elements in our set, one of them is the larger and one is the smaller in our hierarchy? Well, not all elements in partial orders are "relatable" - take the order from above, where a number is "below" another if it's a factor of the former and consider that with the two numbers $3, 5$, neither is a factor of the other and so neither is above or below the other in the hierarchy.

While we're here, can you guess what the Hasse diagrams of wosets look like? The last two axioms make them always look like a line going upwards - to visualise this more easily, a classic example of a well-ordered set is the natural numbers with $\leq$ as the ordering.

Cantor discovered all of this and observed something: since the structure of well-orderings is essentially "linear", it might be possible to define a collection (more specifically, a *proper class*) S of wosets increasing in "length" such that every *possible* woset is order isomorphic to *exactly one* member of S. Cantor

(and subsequently JOHN VON NEUMANN, whose definition we will be using)
called them *the ordinals* and defined S to be those sets that satisfy these
requirements:

1. $\emptyset$ (the empty set) is an ordinal.

2. Ordinals are well-ordered by the subset-relation (i.e. one is "below" another
   if it is a subset of the other).

3. Any subset of an ordinal is an element of that ordinal.

Due to them being order isomorphic, most people notate *finite* ordinals as
the natural numbers, referring to them as the "ordinal numbers" along with
common use of $\leq$ and $<$ in notation instead of the subset symbol, so

$$\emptyset = 0, \{0\} = 1, \{0, 1\} = 2, ...$$

The ordinal that a given well-ordered set is order isomorphic to is called
the *order type* of that set. There are many nice properties that immediately
spring from this definition: a given ordinal *is* the set of all those below it, there
are an infinite number of ordinals and more ... Interestingly, it shows that
there cannot be what Cantor called Ω, *the set of all ordinals*, since such a set
would have an order-type larger than Ω itself since Ω would *be an ordinal* (this
is a contradiction as an element of a woset cannot be strictly above *itself* in
the hierarchy). Cantor, however, was a deeply religious man who connected
Ω philosophically with God as "the absolute infinite", a quantity bigger than
anything conceivable or inconceivable, and so simply concluded that "Ω's nature
is contradictory" and is perhaps "beyond mathematics".

Today, we get around this by saying that there is a "class" of ordinals but
not a set: every property that a mathematical object (like a number) can have
defines a class of objects with that property and every set is a class but not
every class is a set; this removes many paradoxes such as the *Russell Paradox*
posed to set theorists at the end of the 19th century.

There are 3 types of ordinals: 0, successor ordinals and limit ordinals. An
ordinal is a *successor ordinal* if, for some α, it can be represented as the set
$\{x | x \leq \alpha\}$ or, equivalently, $\alpha \cup \{\alpha\}$ (can you see why these are equivalent?).
The intuitive way of thinking about this is that successor ordinals are just $1+$
some other ordinal, e.g. every finite integer greater than 0 is a successor ordinal.
An ordinal λ is a *limit ordinal* if for any ordinal α smaller than it, there is some
other ordinal β that lies between α and λ in the hierarchy - it should soon become
apparent that limit ordinals are what most people would call "infinities"!

Another way of seeing limit ordinals is as upper-bounds to sequences: if
we have a strictly increasing sequence of ordinals that is not finite in length
then we say that the *limit* of that sequence is the smallest ordinal greater than
all those in the sequence. Limit ordinals can thus be seen as the limit of the
sequence formed by all those ordinals below them (the sequence never reaches
the ordinal after a *finite* number of terms but every term is smaller than the
limit). Moreover, one of the key properties of the ordinals that makes them so

interesting to work with is that such a limit *always* exists! Let's examine a few of these limit ordinals.

Consider the sequence $1, 2, 4, 8, 16, 32, ..., 2n, ...$ The limit of this sequence is ω, the first limit ordinal! This number is above all of the integers and, by the definition of ordinals, *is* the set of natural numbers. Cantor called ω a *transfinite number* since although it is above all the finite numbers, it isn't above Ω. There must be a successor cardinal to ω, namely $\omega + 1$ which is a distinct ordering from that of the natural numbers since it has a greatest element (this will be explained in greater detail later when we expound upon the *arithmetic on infinities* that Cantor devised). This too has a successor $\omega + 2$ and so on: $\omega + 3, \omega + 4, ..., \omega + 5$; remember that this is a strictly increasing sequence and so must have a limit, which is $\omega \times 2$.

As you may remember, if we have a well-defined property for ordinals, i.e. it's clear that an ordinal either *does* or *does not* have this property, then we can define a class of ordinals with this property (this is known in modern parlance as the *Axiom of Comprehension*). This means we can define Λ to be the class of all limit ordinals and then, by well-ordering it, we can construct the sequence of limit ordinals in size order. This must have a limit and we can call it λ: the limit of all limit ordinals (which is itself a limit ordinal)!

Ordinals clearly have complex properties but to truly be able to investigate them in depth, we need to consider how one can define functions over the ordinals, just as we can define functions over integers or real numbers. One must begin by understanding *transfinite induction*: to prove a property holds for all ordinals we need to prove it holds for 0 (the so-called base case), we then must show that assuming it holds for an ordinal α then it must hold for $\alpha + 1$ (the inductive step) and finally we must show that for any limit ordinal γ, the fact that all ordinals smaller than γ have the property implies that γ has the property.

Those among you who know about inductive proofs with will be familiar with the first two steps but not the third; it's needed here because limit ordinals aren't the direct successor to any particular number and so the second inductive step doesn't "reach" them.

Now we can completely (inductively) define arithmetic on ordinals. Let's begin with addition:

1. $\alpha + 0 = \alpha$

2. $(\alpha + \beta) + 1 = \alpha + (\beta + 1)$ (the $+1$ denotes the successor ordinal)

3. For a limit ordinal γ, $\alpha + \gamma$ is the limit of the sequence $\alpha, \alpha + 1, \alpha + 2, ..., \alpha + n, ...$ for all $n < \gamma$.

You can think of $\alpha + \beta$ as imagining that α and β were two ordered sets with no common element, $(S_\alpha, \preccurlyeq_\alpha)$ and $(S_\beta, \preccurlyeq_\beta)$, and taking their union $\alpha \cup \beta$ and then imposing a new order on that set $\preccurlyeq_{\alpha + \beta}$ so that the ordering of S_α's and S_β's elements are kept the same and all elements in S_β are above S_α in the ordering (in terms of Hasse diagrams, just imagine drawing the two lines representing S_α and S_β on top of each other to make a diagram of $S_{\alpha+\beta}$). For example, the elements of $2+3$ and their ordering (where $\alpha = 2$ and $\beta = 3$) can be represented in a form of sideways Hasse diagram as

$$0_\alpha < 1_\alpha < 0_\beta < 1_\beta < 2_\beta$$

which is clearly order isomorphic to the ordinal 5, since you can just relabel the elements to get

$$0 < 1 < 2 < 3 < 4$$

This allows us to see that $1 + \omega = \omega$ since

$$0_\alpha < 0_\beta < 1_\beta < 2_\beta < ... \cong 0 < 1 < 2 < 3 < ...$$

(where $\cong$ denotes order isomorphism) but that $\omega + 1$ is something distinct:

$$0_\alpha < 1_\alpha < 2_\alpha < ... < 0_\beta$$

since 0_β comes after all the natural numbers (representing the element ω) and thus represents a greatest element.

Next, we have the definition of multiplication, which is what you might expect:

1. $\alpha \times 0 = 0$

2. $\alpha \times (\beta + 1) = (\alpha \times \beta) + \alpha$

3. For a limit ordinal γ, $\alpha \times \gamma$ is the limit of the sequence $\alpha \times 1, \alpha \times 2, ..., \alpha \times n, ...$ for all $n < \gamma$.

Thus, $\alpha \times \beta$ can be interpreted as taking their cartesian product, which is the set of all *pairings of elements* between α and β, and then imposing an order where an element (a_0, b_0) is below another element (a_1, b_1) if $b_0 < b_1$ or if $b_0 = b_1, a_0 < a_1$. For example, we can display 2×3 in our previous intuitive "order representation" as

$$(0,0) < (1,0) < (0,1) < (1,1) < (0,2) < (1,2)$$

Can you draw out what 3×2 is and why it is order isomorphic to 2×3?

Just like ordinal addition, ordinal multiplication is no longer necessarily commutative (i.e. $\alpha \times \beta$ doesn't always equal $\beta \times \alpha$) when we start dealing with transfinite numbers. For example, we can see from the inductive definition that $2 \times \omega$ is the limit of the sequence $2, 2 \times 2, 2 \times 3, ...$ which is ω but $\omega \times 2 = \omega + \omega$, which can be drawn out as saying that

$$(0,0) < (1,0) < (0,1) < ... \ncong (0,0) < (1,0) < (2,0) < ... < (0,1) < (1,1) < ...$$

It's not necessary to give the inductive definition of ordinal exponentiation here; can you guess what it would be?

From the order representation perspective, we would be simply dealing with longer tuples (the ordered bracket groupings) in our sequences where the length of the tuples is dictated by the exponent and the maximum number that can be counted to in each entry is dictated by the base. So 2^3 can be seen as $(0,0,0) < (0,0,1) < (0,1,0) < (0,1,1) <$ etc.

We now understand the concept of order types fairly well but is there another inherent distinguishing property of wosets that might be important? Size! You may have noticed that I often carefully referred to a situation like $a > b$ as a being "above" b in an ordering, rather than "greater than" or "more than" - this is to distinguish between *ordinality* and size, or what Cantor called *cardinality*. The distinction is not obvious on a finite scale, since the finite ordinal n has n elements, which is written $|n| = n$, but Cantor's great observation was that infinite sets of the same cardinality can have differing orderings. So surely, if we are talking truly about size, then we don't need to worry about all the "bigger infinities" rhetoric - a set is either finite or infinite, right?

Wonderfully, amazingly, it is not so. To prove this, we first observe that one way of proving that two sets A and B have the same cardinality is to exhibit a one-to-one mapping between their elements, known, as we have seen above, as a bijective function, e.g. I can show my hands have the same number of fingers without defining any numbers simply by placing the thumb of my left hand onto the thumb of my right and the index of my left onto the index of my right etc. Further note that a way of proving that $|A| \leq |B|$ is to exhibit a mapping from A to B such that no two elements of A are mapped to the same element of B - such a mapping shall be called *injective*. Then any set S for which there exists an injective mapping from S to the set of natural numbers, ω, we shall call countable and we can call ω the first transfinite countable, since other transfinite ordinals like $\omega \times 2$ and even ω^ω can be shown to be countable.

Let us consider the set of real numbers between 0 and 1; we shall now see that this set is not countable via proof by contradiction. Assuming the set is countable, there must be a one-to-one correspondence between each real between 0 & 1 and a natural number. To visualise this, imagine we have a table with two columns: one column counts through the natural numbers starting at 1 and the other column displays the real number matched to each natural number next to it in the first column. It might look something like this:

1	0.1**0**1010101...
2	0.3**3**4000000...
3	0.27**1**828182...
4	0.314**1**59265...
5	0.1202**0**5690...

Now I shall construct a number ψ which cannot be in the list: to construct ψ's nth decimal digit, one checks the nth digit of the real number assigned to the natural number n - if it is 3, then the nth digit of ψ is 4 and if it is not 3,then the nth digit of ψ is 3. With the example of the table above, we would inspect the digits in bold of each real number and get ψ's value to be 0.34333...

ψ cannot be our list since ψ differs from every number in the list in at least one decimal place. In other words, for any number A you choose from the table, ψ must not equal A because if x is the natural number assigned to A, then we know that the xth digits of A and ψ differ due to ψ's crafty definition... Therefore our initial assumption must be wrong: that the set of reals between 0 and 1 is countable and can thus be put into a table.

This set is often referred to as the continuum (sounds very Dr. Who!) and we say that it is *uncountable* with cardinality "$\mathfrak{c}$". Is the set of rationals countable? Yes, but Cantor's unusual proof of this will left as further reading.

Having seen how some infinities are bigger than others, you might wonder if $\mathfrak{c}$ is, in the grand scheme of things, a relatively large or small infinity and to understand the rather odd situation surrounding that question we must see the formal definition of cardinals, a class of numbers describing the different sizes sets can take:

For any well-ordered set S, we simply define its cardinal number to be the smallest ordinal α for which there exists a bjiective function from the elements of α to the elements of S.

All finite ordinals are their own cardinals so we simply notate the finite cardinals in the same way as we do normally for natural numbers. We notate the transfinite cardinals $\aleph_\alpha$ (pronounced "aleph-alpha") for some ordinal α denoting the αth transfinite cardinality. In particular, the cardinality of the natural numbers is $\aleph_0$ ("aleph-null"), with the ordinal it describes (called its initial ordinal) being ω, and the first uncountable cardinal is $\aleph_1$, whose ordinal is called ω_1.

Arithmetic on cardinals is relatively simple to explain. Regarding addition, $a + b$ for finite cardinals a, b would be evaluated in just the same way as natural number addition and for infinite cardinals a, b it is still true that $a + 0 = a$ but if $a > b$, then $a + b = a$. Similarly, with multiplication, $a \times b$ for finite cardinals would be evaluated as taught in a primary school classroom and for infinite cardinals, while it is true that $a \times 0 = 0$, if $a > b$ then $a \times b = a$. This is because addition is just the union of two sets of cardinalities a, b with no common elements and multiplication is the cartesian product. Exponentiation is less simple to evaluate: for cardinals x, y, $|x|^{|y|} = |x^y|$ where x^y should be evaluated as repeated cartesian products (somewhat similar to ordinal exponentiation).

To demonstrate how evaluating cardinal exponentiation is non-trivial, consider the set operation $P(S)$, described as "taking the power set of S", where it outputs the set of all subsets of S. It can be seen that $|P(S)| = 2^{|S|}$ because for each element of S, you can either include it in a subset or not (2 choices)[2].

So, as Cantor more formally showed, the power set of a cardinal is always bigger than that cardinal. This shows that there are an infinite number of infinite cardinals, since for every $\aleph_n$, $|P(\aleph_n)| > \aleph_n$! Secondly, it allows us to see that the cardinality of the continuum, $\mathfrak{c}$, is equinumerous to the powerset of the natural numbers, $2^{\aleph_0}$.

---

[2]The problem with this is that I'm reasoning on a finite scale only. To prove this equivalence for any set S, you'd have to work more strictly with the above definiton.

Why? Consider the binary representations of all the reals in the continuum - we define the bijective function $f(x)$ from the continuum to the power set of the natural numbers which constructs a unique subset P of ω from x by looking at the nth digit of the binary representation of x and if that digit is 1, then n is put in P, otherwise it's left out. This one-to-one correspondence demonstrates that $\mathfrak{c} = 2^{\aleph_0}$.

So, we've made progress in understanding $\mathfrak{c}$'s cardinality in reference to the other aleph numbers, right? NO - a conjecture Cantor made was that

$$\mathfrak{c} = \aleph_1$$

This is called the *Continuum Hypothesis* and in 1940, the great logician KURT GÖDEL showed that the axioms of ZFC set theory are not capable of *disproving it*, and in 1966, the great set theorist PAUL COHEN won the Fields Medal (the equivalent of the Nobel Prize in mathematics) for showing that ZFC set theory is not capable of *proving it*. What the "answer" is to the Continuum Hypothesis perhaps has no meaning - an extremely scary idea in a subject like mathematics where statements are supposed to be either *true* or *false* - and to continue working, one must either accept it or deny it (i.e. one can add axioms which allow its proof or disproof and in both cases retain consistency). One can assume that $\mathfrak{c} = \aleph_2$ or $\mathfrak{c} = \aleph_{23}$ without leading to contradiction; in fact the first cardinal for which one cannot "safely" assume a variant of the Continuum Hypothesis is $\aleph_\omega$.

We're sure you would agree that all this work (of which Cantor did much more) is quite incredible but unfortunately not everyone historically has agreed. Many contemporaries of his criticised him and his work, perhaps most famously when the number theorist Leopold Krönecker said "God made the integers, all else is the work of man." and "I am sure there is no mathematics there [referring to the theory of infinite sets]". As a result, Cantor's life was full of grief and depression, leading to early hospitalisation and death. We hope his work may live on in tribute to a great mind.

For further reading on this topic, try:

1. Beth Numbers

2. The amazing Fixed Point Lemma for continuous ordinal functions

3. The Church-Kleene Ordinal and other large countable ordinals

4. Infinite Chess Ordinals

This article was named after a very young online wiki - for those who continue to read in this subject, please add to it!

Challenge IV:

1. Evaluate $\omega \times \omega^{\omega+20}$

2. Many ordinals can be expressed in the form $\omega^{a_0} \times b_0 + \omega^{a_1} \times b_1 + \omega^{a_2} \times b_2 + ...$ where $a_0 > a_1 > a_2 > ... \geq 0$ and all a_i are finite ordinals. This way of

notating such ordinals is known as the *Cantor Normal Form* and can be seen as analogous to a base-ω numerical notation system. Is there an ordinal for which there does not exist a finite representation of it in Cantor Normal Form? (HINT: There *is*, but is it countable? Can you describe it? Are there others?)

Chapter 5

Geometric Dreams

Salutations!

Geometry problems are a staple of modern day maths tests, qualifications and olympiads - indeed, EUCLID is considered perhaps the first modern mathematician and his **Elements**, used as a textbook for hundreds of years, predominantly focusses on geometric constructions on the plane. In fact, for many mathematicians in the 19th century, geometry in many dimensions was almost considered "solved", in that there were methods which could eventually given enough time understand all but a few objects. In some sense, geometry was the mathematical realisation of the real world for them and thus was a tool for physicists to describe and predict various aspects of it - they could never have predicted that there wasn't just *one geometry* and that many of these other geometries would have remarkable properties completely unlike that of euclidean space. This article will mention the profound discovery of non-euclidean geometries and focus on one particularly, known as *hyperbolic geometry*, with some showcasing of even wilder beasts such as the quintic 3-folds of the complex projective space near the end...

Numerous people discovered hyperbolic geometry at around the same time, though interestingly in quite different contexts. One of the main trends of the era was to look into building a proper and rigorous foundation for contemporary mathematics for it had been discovered that without exact definitions, clear axioms and systematic work of this sort one would soon find that mathematical questions could be asked that had no answer, simply because the concepts were too vague and so were not well-defined in all circumstances. The Hungarian mathematics student JANÓS BOLYAI was also interested in just this, looking particularly at Euclid's foundations of geometry. As many scholars of the time would tell you, Euclid put forth 5 main axioms:

- Given any two points, a straight line can be drawn connecting them.

- A straight line can be continuously extended in either direction unboundedly.

- All right angles are equivalent to each other.

- Given a centre and a distance, a circle with that centre and with a radius of the given distance can be constructed.

- Given a line A and a point B not on the line, there is exactly one new line C that can be drawn through the point so that C does not ever intersect A.

If any of the above seems quite obvious to you, then that's good news! An *axiom* is supposed to be a principle or truth that is considered so obvious that it does not require any proof; as such, mathematical theories or objects are typically begun with a list of such axioms making clear what is the topic of discussion. Euclid then went on in his book to describe various different geometric situations on the plane and prove fundamental theorems used in geometry today, such as the fact that the interior angles of a triangle sum to 180°. However, what became clear to many reading this prized document was that in some sense, the fifth axiom or *parallel postulate*, as it has come to be known, was significantly detached from the first four. This is because Euclid's first 25 or so theorems never used the fifth axiom whereas the others were used extensively - many wondered if it was actually possible to prove the fifth from the first four. When Bolyai read about this, he realised its importance and decided instead to attempt to prove that fifth axiom *had* to be how it was because if it was any different, one would get an inconsistent (and thus, invalid) geometry; in other words, he tried a proof by contradiction. Amongst other things, he used this replacement of the parallel postulate, which appears at first glance to be faintly ridiculous:

- Given a line A and a point B not on the line, there are two lines C_1 and C_2 that can be drawn through the point so that neither ever intersect with A.

So, Bolyai decided to start investigation this "non-geometry" and noticed amongst other things that triangles in this odd world can only have a sum of internal angles less than or equal to 180°. After this, he showed that in a quadrilateral two equal-length sides A and B perpendicular to another side C the other two angles would be acute. As he went on proving theorem after theorem, it eventually occurred to Bolyai that in fact there was no a contradiction to be found - he and others after him succeeded in proving, astoundingly, that this new "hyperbolic" geometry was consistent if and only if Euclidean geometry was consistent (a belief commonly held by contemporary mathematicians).

Some of his other founding work included the description of the trigonometric functions for right-angled hyperbolic triangles $sinh(x)$, $cosh(x)$ and $tanh(x)$ and proving some elementary identities concerning these. Furthermore, he showed that the "opposite" of his replacement of the parallel postulate, namely

- Given a line A and a point B not on the line, there are no lines that can be drawn through the point so that they never intersect with A.

also defines a consistent geometry, called "elliptic geometry". On the elliptic plane, triangles always have *more* than 180° in their sum of internal angles and tilings can be made such that a finite number of tiles cover the entire plane. It turns out that the elliptic plane is easily describable as the geometry on the surface of a sphere, where we represent a "point" as two opposite points (known as *antipodal* points) on it and a line is a great circle of the sphere[1] - so one of the tilings I mentioned can be seen as in Figure 5.1, which is, by the way, analogous to the icosahedron.

Contemporaries of Bolyai, such as LOBACHEVSKY and even JOHANNES KARL FRIEDRICH GAUSS himself, similarly independently discovered hyperbolic planes

---

[1] A great circle on a sphere is one of the circles that one could draw on its surface such that the radius of the circle is equivalent to the radius of the sphere.

Figure 5.1: A tiling of the elliptic plane.

and geometry but through a different series of thoughts. It had occurred to them that on different surfaces from the flat plane one could also have a different geometry but classifying it was not as simple as it might appear. For example, a cylinder appears different from the plane however its surface has no distinct geometric properties; this can be seen by the fact that any figure or diagram on a euclidean plane can be put on a cylinder preserving angles and lengths by picking up the plane and wrapping it around the cylinder like wallpaper (for example, toilet paper being on a roll).

However, it is *not* true that one can easily wrap a piece of paper around a sphere - one always finds wrinkles when doing so. Why is this? Gauss forumulated the concept of Gaussian curvature to understand this and to better aid with his cartography of the Austrian Alps.

The curvature of a plane curve[2] at a specific point A can be considered intuitively as a measure of how effective a tangent at A can be used to approximate the location of points just around A or how fast the tangent diverges from the actual curve as one moves away from A. More rigourously, given any curve and a point A somewhere along it, there is a circle with a centre O on the line perpendicular to the tangent at A (called the *normal*) with $|OA| = r$ for some r (r is the radius). This is the best circle which approximates the curve for points close to A - an illustration of this concept can be seen in Figure 5.2. We define κ to be the curvature, where $\kappa = \frac{1}{r}$.

Using this, we can define Gaussian curvature, which is a way of measuring the intrinsic curvature of surfaces. Given a point A on a surface, consider the normal at A (here this is the perpendicular line to the *tangent plane* at A) and then, further, consider all the planes that contain that normal. Each of the

---

[2]A plane curve can for our purposes be considered any line, wiggly or straight, drawn continuously on a surface.

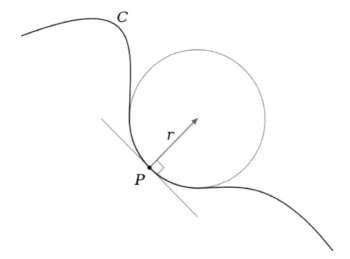

Figure 5.2: A diagram of curvature at a point P with its "approximation circle" of radius r.

planes will intersect with a 1-dimensional cross-section of our surface and so for each intersecting plane P_n, we calculate the curvature of the corresponding 1-dimensional plane curve at A, denoted κ_n. Now, the *Gaussian curvature* of A is defined to be the product of the maximum and minimum such κ_n (known as the *principal curvatures* at the point A). So, let's stop to consider what the Gaussian curvature of various well-known surfaces looks like:

- with a point on a plane, the curvature of a point on any plane-cross-section is 0, since the best circle which approximates a straight line has an infinitely large radius[3], so $\kappa_{max}, \kappa_{min} = 0$ so the Gaussian curvature is 0;

- with a point on a cylinder, the curvature of the long straight axis is κ_{min} which is 0 for the same reason as on the plane and so regardless of κ_{max}, its Gaussian curvature is 0;

- with a point on a sphere, every plane-cross-section will be a circle of radius R, meaning that $\kappa_{min}, \kappa_{max} = \frac{1}{R}$ and so the Gaussian curvature is $\frac{1}{R^2}$ at every point.

What would a point of negative curvature on a surface look like? Consider Figure 5.3, displaying a so-called "saddle-surface". At the centre of the saddle, the diagram displays the planes which cross-sect the surface at its principal curvatures - one of the cross-sections curves upwards in the direction of the normal and the other, 90° to the first, curves downwards against the direction

---

[3] In other words, $\lim_{R \to \infty} \frac{1}{R} = 0$

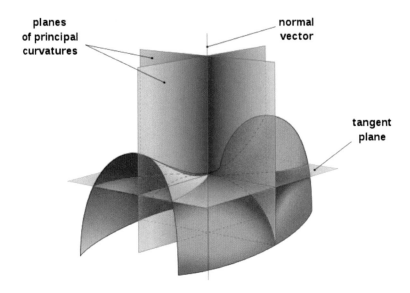

Figure 5.3: A "saddle-surface".

of the normal. The former would have some positive κ_{max} but since the other curves downwards from the normal, Gauss said it would be natural to assign the downwards curve a negative κ_{min} and so the Gaussian curvature (their product) would also be negative. The interesting question that Gauss asked after considering this was what would a "plane" with constant negative curvature look like? After giving the problem some thought, one might come up with the surfaces shown in Figure 5.4; one is a *hyperboloid* (the surface of revolution of a hyperbola) and the other is called a *pseudosphere*, discovered by Bolyai. Both have constant Gaussian negative curvature but the pseudosphere has a circle of edges near its center at which curvature is undefined and the hyperboloid has the problem that circles around the z-axis are different sizes depending on their height and so neither case constitutes a "plane" like the circle or the 2D euclidean space, meaning that we couldn't create, for example, a tiling of either while making sure that every tile has the same sidelengths and the angles of each prototile are kept the same.

In trying to understand the problem better, one can examine the geometric properties of such a surface and what one discovers is that triangles have less than or equal to 180° as their summed internal angles etc. - this is just a different description of the hyperbolic plane! Although Gauss never did gain much better insight into the nature of such a surface, he did prove the important *Theorema Egregium* which says that by performing cutting, rotating, translating or bending operations on a surface, one sustains the same Gaussian curvature at every point; thus, Gaussian curvature is a defining "unalterable" property of a surface, showing clearly why one can't wrap a piece of paper around a sphere as easily as around a cylinder - the piece of paper has zero constant

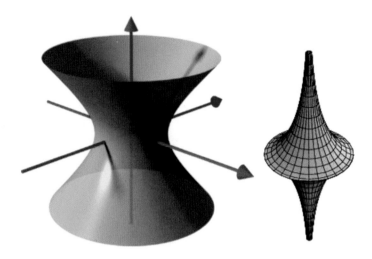

Figure 5.4: A hyperboloid and a pseudosphere.

curvature and the sphere has $\frac{1}{R^2}$ and so to cover the sphere with paper one would have to be able to stretch or otherwise fundamentally alter its nature. It also shows why rolling up a piece of paper into a cylinder makes it so much more rigid than when it was just flat: when you've rolled it up, you have created a positive curvature in one direction and so for the paper to retain its constant zero Gaussian curvature, it has to stay straight in the other direction...

The final piece of the puzzle came from the work of the prolific David Hilbert with *Hilbert's Theorem*[4], which essentially proved that given an n-dimensional hyperbolic geometric space (as in, an n-space of constant negative curvature) there is no distance and angle-preserving way of putting it into an n-dimensional Euclidean space, certifying that this is a consistent and fundamentally distinct geometry.

For those who read the first article in this series, you would have seen an introduction to some of the many combinatorial and geometric problems and properties of tilings of the Euclidean plane. It was suggested then that hyperbolic geometry sported a much greater collection of such tilings and now we shall see that such comments are quite justified and that, in some sense, most tilings are in fact hyperbolic.

The *Schläfli symbol* $\{x_2, x_1\}$ is a notation used to denote a tiling where

---

[4]It should be noted that this name, just like *Euler's Theorem* in modular arithmetic, is completely ridiculous since its namesake proved many hundreds of theorems and is commonly considered to be amongst the most important of mathematicians of the 19th and 20th centuries.

there are x_1 x_2-gons around a point. So, for example, the Schläfli symbol $\{4,4\}$ describes the classic square-lattice tiling on the Euclidean plane and $\{3,5\}$ describes the tiling in Figure 5.1. You might notice that this notation severely limits the number of tilings one can describe since you can only notate those which use only 1 shape and in which each vertex has the same number of polygons around it (these are known as the *regular tilings*) but it is still useful. What symbols $\{p,q\}$ are Euclidean?

Well, it can only be Euclidean if the sum of q of the internal angles of p-gons is $360°$, that is $q(180° - \frac{360°}{p}) = 360°$. This is equivalent to the statement that

$$(p-2)(q-2) = 4$$

By graphing this line with one axis being p and the other q, we get Figure 5.5.

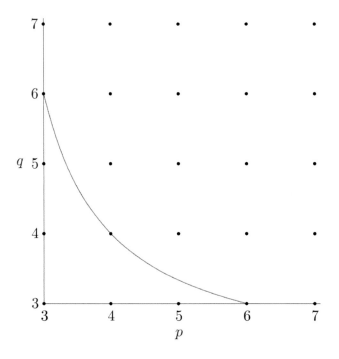

Figure 5.5: A graph of the Euclidean tilings.

This diagram is quite powerful in that it shows clearly the space of regular tilings in each geometry. Those on the line are Euclidean such as $\{3,6\}$, those below the line are elliptic such as $\{3,4\}$ and the infinity of those above it are *all* hyperbolic! With so many, one would hope that there would be a way of visualising or creating images of these tilings - thanks to the work of HENRI POINCARÉ and FELIX KLEIN, you can look at visualisations of the hyperbolic plane.

These visualisations act as a projection of the entire hyperbolic plane to the unit disk[5] where the edge of the disk can be seen as the "edge at infinity" of the hyperbolic plane - although I will not mention the specifics of geometric constructions or calculations with them in this article, it should be clear that distances in the disk going out from the centre become increasingly large for the corresponding hyperbolic figures. In the Poincaré disk model, lines are represented by circle arcs drawn in the disk which are orthogonal to the disk's edge and the model represents angles accurately (this property of a mapping from one space to another is known as *conformality*). In the Klein model, conformality is lost with the benefit that lines are drawn straight here. The same $\{7, 3\}$ tiling is represented using the two models in Figure 5.6 - though in general the images I will use in this article and any articles involving hyperbolic space in the future will use the Poincaré disk model due to its angle preserving nature.

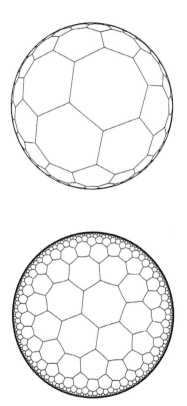

Figure 5.6: A $\{7, 3\}$ tiling projected into the Klein Model and the Poincaré Model.

---

[5]The set of points on the Euclidean plane of distance less than or equal to 1 from $(0, 0)$.

Another incredible object that can exist in hyperbolic tilings is the *apeirogon* or ∞-gon (a polygon with an infinite number of sides). In Euclidean tilings the only way it can be represented is in "degenerate"[6] tilings, such as can be found in Figure 5.7, where the "sides" of the polygon are just an infinitely long line separated by dots denoting vertices at regular intervals. The image can be seen as a tiling of the Euclidean plane by 2 apeirogons.

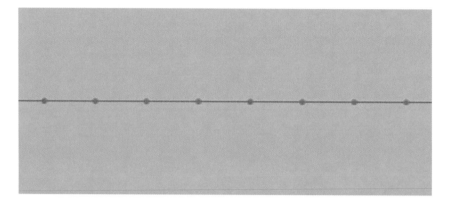

Figure 5.7: A $\{\infty, 2\}$ tiling.

However, hyperbolic space is sufficiently curved and "large" that apeirogons can exist as discrete shapes surrounded by other polygons and that, equally, there can be tilings with an infinite number of polygons around each vertex, such as those shown in Figure 5.8! In this case, all the vertices of the tiling are actually points at the edge of the disk and thus at infinity, known as *ideal points*.

Hyperbolic geometry becomes even more complicated in 3-space, or 4 or further! In order to investigate these spaces we need to understand the extended definition of the Schläfli symbol and learn about *horocycles*. In general, the Schläfli symbol $\{x_n, x_{n-1}, ..., x_2, x_1\}$ denotes x_1 copies of $\{x_n, x_{n-1}, ..., x_2\}$ around each vertex - this allows us to talk about regular *honeycombs* of arbitrary dimensions! You can see some examples of different cube-based polyhedron packings in Figure 5.9, going from the elliptic to the hyperbolic (represented here by the Poincaré *sphere* model).

Most interesting, however, is the next in the series, namely $\{4, 3, 6\}$, whose vertices are purely ideal. To see why $\{4, 3, 6\}$ would suddenly have only ideal

---

[6]The word *degenerate* is used in mathematics to describe objects or situations which technically fulfill specific requirements but are so devoid of structure or meaning that they are worthless to study. Another example of a degenerate object would be the 2-sided or 1-sided polygons (which, on the Euclidean plane, are just straight lines).

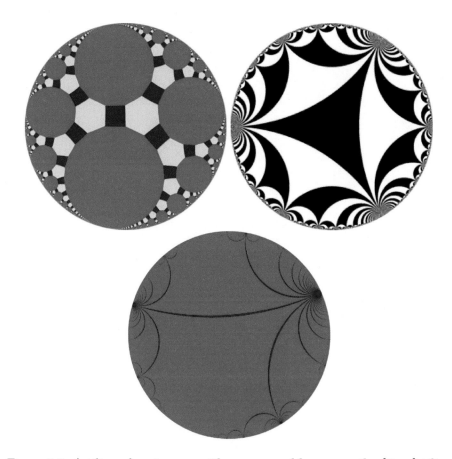

Figure 5.8: A tiling of apeirogons with squares and hexagons, the $\{3, \infty\}$ tiling and the $\{\infty, \infty\}$ tiling.

vertices while all its predecessors have none, we can look at its *dual honeycomb*. The dual of a honeycomb is formed by taking the geometric centres of every face of every polyhedron involved and drawing lines between those that are adjacent to each other (as in, those which are part of a common polyhedron). 2D examples are easier to visualise at first, so let's take a look at some: in Figure 5.10, we see two diagrams, one showing how the square tiling $\{4, 4\}$ is self-dual, where the original tiling is in blue and the dual is in red, the second showing how the dual of the cube (corresponding to the Schläfli symbol $\{4, 3\}$) is the octahedron (with symbol $\{3, 4\}$. Can you spot a pattern here?

The dual of a geometric structure defined by the symbol $\{x_n, x_{n-1}, ..., x_2, x_1\}$ is $\{x_1, x_2, ..., x_{n-1}, x_n\}$! Can you think why this is?

It allows us to see that the dual of $\{4, 3, 6\}$ is $\{6, 3, 4\}$, which can be interpreted as having 4 of the polyhedron $\{6, 3\}$ around a point... except that $\{6, 3\}$ defines the normal Euclidean hexagonal tiling of hexagons... Remarkably, hyperbolic space is so curved that it permits Euclidean tilings as infinite-sided polyhedra

(A) $\{4,3,3\}$ (B) $\{4,3,4\}$ (C) $\{4,3,5\}$

Figure 5.9: A series of three poyhedron packings in different spaces with their respective Schläfli symbols.

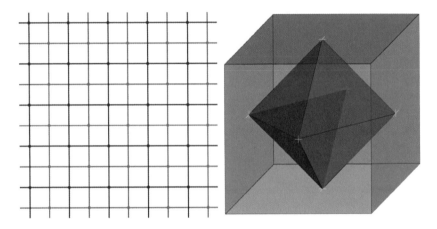

Figure 5.10: The duals of $\{4,4\}$ and $\{4,3\}$.

where the plane curves in on itself and becomes a form of hyperbolic ball - such an object is known as a *horocycle*[7] and to me it's simply astounding that such an object could exist. Consequently, since the definition of a dual honeycomb B of a honeycomb A requires that the number of faces of the polyhedra in B is the same as the number of the edges going into the vertices of A, it follows that there are infinite number of lines going into the vertices of $\{4,3,6\}$ and thus its vertices are ideal.

The mathematician Felix Klein started modern geometry by founding his *Erlangen* program[8]. He said that we could start with a geometric space that we

---

[7]The official definition of a horocycle is a curve in hyperbolic space with the property that all lines intersecting it orthogonally tend to the same ideal point, but for our purposes the above commentary will suffice.

[8]An interesting fact the author noticed about Klein is that his birthdate was 25/04/1849 - so the day, month and year were all square numbers! He must have been destined to be a

were familiar with, such as 3D euclidean space, and then impose symmetries of various kinds to it to create a *new geometry* to be studied and that by choosing our symmetries carefully we could tailor our geometries to be relevant to certain mathematical problems so that by examining the geometry, we could gain insight into those problems. That was quite abstract, so to illustrate what we mean by *symmetries*, let's take a concrete example:

Consider the Euclidean plane $\mathbb{R}^2$ - the squared notation above the real-numbers symbol here represents "the set of all pairs of reals". Now, our symmetries are a collection of operations that we can perform on these points with the idea that if you can get from point A to point B via one of the operations then in our "new geometry" they will be considered the *same point*. If we said, for example, that our new symmetries on $\mathbb{R}^2$ were the operations of adding 1 to the x value of a point and adding 1 to the y value of a point, then we essentially collapse the entire plane into just the unit square (the set of points whose x and y values fall between 0 and 1) - can you see why? This square would be just like the game *Pacman* because if you travel up the y-axis to the edge of the square and go beyond the top edge, you end up at the bottom of the square and the corresponding fact is true for going along the x-axis! So, in a strange sense, we have collapsed our plane into *a torus* (or doughnut) because if you "go up" in a torus, then you get back to where you were, and the same is true of "going" left or right! Similarly, if we were to say that every point in the unit square of the form $(x, 0)$ was the same as $(x, 1)$ and everything of the form $(0, y)$ was the same as $(0, 1 - y)$ (you can imagine this as "sealing together" the top and bottom edges and twisting the sides before sealing them) then you get a remarkable surface called a *Klein bottle* (named after Klein, who came up with it through just these ideas) - both of them are shown in Figure 5.11.

The Klein bottle is so bizarre that a model of it cannot be constructed in 3D euclidean space without intersecting itself and has no inside or outside! The kind of modern-day geometry that mathematicians' study is even stranger than much of what we have talked about here. Instead of studying simply the Euclidean plane, one studies general *affine planes*, for which Euclid's first and fifth axiom hold along with the requirement:

- There are four points such that no line is incident with more than two of them.

and instead of studying the elliptic plane, mathematicians examine *projective planes*, in which Euclid's first axiom holds, there are no parallel lines and the above requirement holds. Further, both of the planes just mentioned can be realised using any number system (or for those who know some abstract algebra, any *field*) - for example, one can create the affamed *Complex Projective Plane* by considering $\mathbb{C}^2$ and then calling the set of lines going through the origin "points" or, equivalently, saying that two points (x_1, x_2, x_3) and (y_1, y_2, y_3), with $x_n, y_n \in \mathbb{C}$, are "the same" if there is some complex number z so that $(zx_1, zx_2, zx_3) = (y_1, y_2, y_3)$. In this odd world where points' coordinates are

---

mathematician.

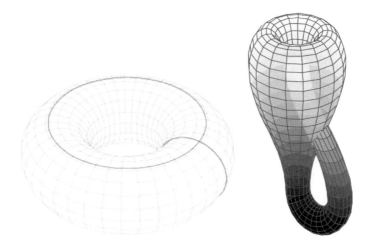

Figure 5.11: *Left*, a torus with the circles highlighted illustrating its periodicity properties on going "up" and "right", *right*, a Klein bottle.

described using complex numbers and lines are planes intersecting the origin, one can have many objects that we could never have in our own geometries such as the 3-dimensional surfaces known as *3-folds* - these exist in Complex Projective 4-space and often have so many rotational or complex symmetries that they simply have no analogue in our universe. For many of these objects, they are so wonderfully complicated that there is no *possible* diagram I could draw to show you what they look like.

For those of you that have enjoyed this article, you may wish to read up on *differential geometry* in which *Riemannian manifolds*, a generalisation of traditional geometric surfaces to any number of dimensions and which only have the requirement that areas "near" each point act analogously to Euclidean space[9] - it has really allowed mathematicians to better grasp what "spaces" *are* and the various properties that they could have... Riemann himself was the top student of Gauss and formulated the above concepts after being introduced to Gaussian curvature by his mentor! So, for further reading, look up:

- The notion of a *field* in abstract algebra and how they are used in the definitions of *projective* and *affine planes.*

- *Topological* spaces.

- *Uniform tilings* of the hyperbolic plane; there's an excellent Wikipedia page on this that contains many more images than in this article!

---

[9]The official definition is a *topological space* which has a neighbourhood associated with every point that is *homeomorphic* to some Euclidean *n*-space equipped with a useful object for measuring distances known as a *Riemannian metric*, but that's for extra reading!

- *Riemannian* geometry; although you shouldn't expect to pick this up immediately, since it is a complex subject, true understanding of this should make General Relativity easy to read!

- Hypercycles and the actual definition of horocycles

- Definitely, if nothing else, go to "h3.hypernom.com" and try and use your arrow keys and WASD to move around - it's a simulator of hyperbolic 3-space from within the Poincaré sphere.

Also, I suggest you look up alternative ways to construct a Klein bottle, one of which is suggested by the following limerick:

There was a mathematician named Klein,
Who thought the Möbius strip was divine,
Said he, "If you glue
the edges of two,
You'll get a weird bottle, like mine!"

And now, as per usual the challenge with a relatively simple first part and a second part that I know *no* good solutions to!

Challenge V:

1. What structure does the symbol $\{5, 3, 4\}$ describe?

2. Can you come up with a notation for tilings, polyhedra and honeycombs which works for *non-regular, non-vertex-transitive* structures?

Chapter 6

Numbers that are *not* astronomical in size

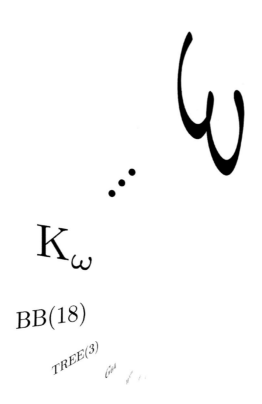

Salutations!

This time we will examine some of mathematics' largest numbers; I feel this is a conceptual area that many find interesting, and even amusing, to think about - some of the numbers I will mention here will be so large that to refer to them as *astronomical* would not just be inaccurate, since there is no object that could found in these quantities within the observable universe, but, frankly, *insulting* to their magnitude.

To demonstrate the previous point, we shall begin by considering the number of baryons in the observable universe. Baryons are particles made up of 3 quarks and interact with the strong nuclear force, e.g. protons or neutrons, and we can calculate how many there are using 4 numbers, 3 of which were obtained using data from the Planck Satellite:

- ρ_{crit}, the critical density of the universe $(= 8.64 \times 10^{-33} kg m^{-3})$

- Ω_b, the fraction of the universe's energy in baryons $(= 0.0485)$

- L, the radius of the observable universe, which is roughly spherical $(= 4.39 \times 10^{26} cm)$

- m_p, the mass of one proton $(= 1.67 \times 10^{-27} kg)$

Now, since ρ_{crit} is essentially the energy density of the universe, $\rho_{crit} \times \Omega_b$ is the mass stored in baryons per cm^3 of the observable universe on average, making $\rho_{crit} \times \Omega_b \times \frac{4}{3}\pi L^3$ roughly the combined mass of all baryons in the universe. Finally, since a neutron's mass is essentially equivalent to that of a proton, we divide the above expression by m_p to get

$$\frac{\rho_{crit} \times \Omega_b \times \frac{4}{3}\pi L^3}{m_p} = 8.89 \times 10^{79}$$

which is really quite a big number, in comparison to the numbers of things you encounter for everyday life! However, it was small enough to be expressed, to a fair level of precision and concisely, using a notation we are so familiar with that I barely need to name it: that of the *exponential*. For many, if asked to write down quickly the biggest number they could think of at the time, exponentials or stacked exponentials (those of the form $a^{b^{c^{d^{\cdots}}}}$) would be their first thought, since it's so simple - for example, just 10^{10^2} is bigger than the number of baryons in the universe. In fact, our first famous number can be expressed as 10^{100}, a *googol*, and the next as $10^{10^{100}}$, a *googolplex*. We shall return to exponentials and the process of stacking them later, for it has great potential to make large numbers.

For now, we take ourselves back to near the beginning of the 20th century, where individuals such as KURT GÖDEL, ALAN TURING and ALONZO CHURCH were discussing the nature of functions. They realised that the process of calculating the outputs to most functions could be seen as an iterative process that, most importantly, had a predictable number of steps; for example, to calculate $2 + 2$, one could see it as applying $f(n) = n + 1$ to the input 2 twice. Such functions were called *primitive recursive*, because they *could* be written down or represented recursively, i.e. where they were seen as a series of repeated applications of some function, but could also be written down in a single closed form - all polynomials, exponentials and many more that we are familiar with are primitive recursive. The computer scientist ROBERT ACKERMANN is most famous for describing an eponymous function, denoted $A(m, n)$, that it was still possible to evaluate but was not primitive recursive defined by these conditions:

$$A(m, n) = \begin{cases} n + 1 & \text{if } m = 0 \\ A(m - 1, 1) & \text{if } m > 0 \text{ and } n = 0 \\ A(m - 1, A(m, n - 1)) & \text{if } m > 0 \text{ and } n > 0 \end{cases}$$

Let us call a *closed-form* representation of a function a form which uses a finite number of operations and without self reference. Then, an amazing fact is that the Ackermann function's above self-referential or *recursive* definition cannot be written out into a closed form, unlike addition or multiplication - this is what it means for it to not be a primitve-recursive function and it grows extremely quickly - try evaluating it for different inputs! Clearly things like $A(0, 3) = 4$ and $A(1, 2) = 4$ are quite small, but then $A(4, 3)$ is an incredible 19729 digit number, roughly equivalent to $2^{2^{65536}} - 3$. In fact, it's often difficult to find examples to demonstrate how large the numbers are that the Ackermann function outputs, because nearly all of them are so big that they either can't be written down in any concise manner or, worse, they couldn't be computed within the lifetime of the universe given all the computing available today. Furthermore, Ackermann and his peers were later able to show that functions of this kind[1] *dominate* all primitive recursive functions, i.e. for any primitive recursive function $f(x)$ and a non-primitive-recursive function $g(x)$, there is some input n so that for all $m > n$, $g(m) > f(m)$.

In order to understand and express just *how* quickly such functions grow, we have to use a lovely typographical system developed some years ago by the famous DONALD KNUTH[2] known as *up-arrow notation*, which is based on the idea of the *hyperoperation hierarchy*. The first operator in the hierarchy is that of the *successor*, an unary operator (meaning that it takes 1 argument) which takes in n and outputs $n + 1$, often written $n + +$.[3] Addition can be seen as repeated successorship in that $a + b$ can be seen as denoting $a + +$, b times.

---

[1] As in, those that can be evaluated in a finite amount of time but that are not primitive recursive.

[2] A computer scientist and mathematician, perhaps most famous for his remarkably complicated series of volumes *The Art of Computer Programming* (often referred to as the computer scientist's bible!) but also for the typesetting system *TeX* which this very publication uses to format its articles!

[3] This is, interestingly, why C++ is called what it is - it was supposed to be the *successor to C*

Multiplication can then be seen as repeated addition in that $a \times b$ represents $a + a$, b times. This continues as we go higher up the hierarchy, with each nth operation $a *_n b$ representing performing the $(n - 1)$th operation to a by itself, b times. Knuth created the hyperoperation-notation $a \uparrow b$ which *starts* at exponentiation (as in $a \uparrow b = a^b$) and by writing more arrows, one goes up the hierarchy, so $2 \uparrow\uparrow 4 = 2^{2^{2^2}}$ - the name we give for this operation above exponentiation is "tetration" and $a \uparrow\uparrow\uparrow b$ is called "*a pentated by b*" etc. These operations make writing really large numbers simple and if we *index* the arrows, that is say that $\uparrow^n$ denotes n arrows, then we can write down numbers that could never have any practical use - for example, the famous *Graham's number*.

This number comes out of a question in a somewhat ill-defined area of mathematics known as Ramsey Theory, which purports to comprehend the conditions under which complex structures are forced to appear; Ronald Graham and Bruce Lee Rothschild, both legends in this field, came up with the question in 1970. The question requires understanding what a *graph* is in pure mathematics; a summary is that any set of *points* and *curves* drawn to connect them is a graph. More formally, a graph is a set of points along with a set of pairings defining connections between those points - thus neither the precise coordinate/relative position of points nor the shape of the lines connecting them matters, only the connections[4].

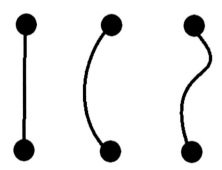

Figure 6.1: These three 2-point graphs are the same.

Given n points, the graph obtained by adding all possible connections between them is called the *complete graph on n vertices*, denoted K_n (e.g. K_3 is like a

---

[4]To be precise, we consider two graphs that have the same number of vertices and the same connections between those vertices but are drawn differently to be distinct graphs or objects but we say they are *isomorphic*, i.e. share all the same graph-theoretic properties.

triangle and K_4 is like a square with its diagonals drawn in). Now, Rothschild and Graham were considering complete graphs of n-dimensional cubes, which have 2^n vertices each, and properties of the *colourings* of their edges, i.e. the ways in which you can assign different colours to those edges. In particular, they asked what was the smallest value of n such that every 2-colour colouring, using, for example, red and blue, of the edges of the complete graph on an n-dimensional cube is *forced* to contain a subset S of its points such that all the edges between the points in S are the same colour and such that all points in S are *coplanar*[5]. They were able to prove that there is such an n and they knew from checking on paper that $n > 5$ and so they sought to also put an upper-bound on it (Graham's number)[6]. It is constructed as follows:

- Let $G_1 = 3 \uparrow^4 3$ (an amazingly large number, so big that the number of 3s in its power-tower representation couldn't be written in base 10 even if each digit could be assigned to each planck-volume in the observable universe!)

- For each n, let $G_{n+1} = 3 \uparrow^{G_n} 3$

- Then Graham's number is G_{64}.

Can you see how large it is? We urge you to ponder over this for a minute, since there are no metaphors or real world analogies for this...

It is clear from this that uparrow notation becomes inadequate for integers as large as Graham's Number, since there is no way of expressing it concisely if we need to write out all the arrows. Thus, when you have gotten over G_{64}, we must move on to a better framework that will allow us to see just how large it is "in the grand scheme of things".

The fast-growing hierarchy is a series of functions, built recursively, that grow faster and faster as we go up. We start with the simple function $f_0(x) := x + 1$ and we say[7] that $f_1(x) := f_0^x(x)$, or in other words $x + x$. Similarly, $f_2(x) := f_1^x(x) = x \times x$ and in general for any integer $n > 0$, $f_n(x) = f_{n-1}^x(x)$.

So far, there is no difference between this and hyperoperations but now, we can use *ordinals* to give us unbounded growth-rates... There was a previous article introducing readers to the wonderful universe of ordinals in **Chapter 4: Cantor's Attic** of this book but, to simplify their technical definition, they are a clever set-theoretic version of numbers, discovered by GEORG CANTOR, which essentially allows us to have a natural extension of the integers to varying sizes of infinity. The number ω is the ordinal "larger" than all the integers but then

---

[5] i.e. are points on a common plane.

[6] It may be of interest that subsequently we have created a better bound, $2 \uparrow\uparrow\uparrow 6$.

[7] Here, $g^n(x)$, for some integer n and some function $g(x)$, denotes performing g to the input x, n times.

we still have a well-defined concept of $\omega + 1$ or $+2$ or $+n$ and much, much more. We call ω the first *limit ordinal*, meaning that it has no specific predecessor, but rather can be reached as a limit of a strictly increasing sequence, and we call $2, 3, 4, n, ...$ and $\omega + 1, \omega + n$ etc. *successor ordinals* because they *do* have a well-defined predecessor (i.e. they are the successor of some known ordinal). Thus we have the definition that if α is a successor ordinal, then $f_\alpha(x) = f_{\alpha-1}^x(x)$, and if α is a limit ordinal and S_α is a strictly-increasing sequence of ordinals whose limit is α (as in, α is the smallest upper-bound for all the terms in S_α), with $S_{\alpha[n]}$ denoting the nth term of S_α for some ordinal n, then $f_\alpha(x) = f_{S_{\alpha[x]}}(x)$.

To give an example[8], $f_\omega(x) = f_x(x)$, since the sequence of integers $1,2,3,...,x,...$ has the limit ω but since $\omega + 1$ is a successor ordinal, $f_{\omega+1} = f_\omega^x(x)$. We can observe from these definitions immediately that $f_\omega(x)$ can't be primitive-recursive, since it grows faster than any f_n for integer n, and thus that it is, in a sense, *beyond uparrows*, since it can't be represented in the form $m \uparrow^k x$, where m, k are fixed integers. In fact, it is possible to show that $f_\omega(x)$ grows at almost exactly the same rate as the Ackermann function that we've seen previously and that $f_{\omega+1}(64) > G_{64}$.[9] Now, you can choose your favourite transfinite ordinal and create a function that grows faster than you can imagine, for example $f_{\omega \times 2}, f_{\omega^2}, f_{\omega^\omega}$ or, if $\epsilon_0 = \omega^{\omega^{\omega^{\cdots}}}$, then you can even have f_{ϵ_0} and larger!

Kruskal's Tree Theorem, conjectured by ANDREW VAZSONYI and proved in 1960 by JOSEPH KRUSKAL (an influential combinatoricist), is a statement, once again, relating to graphs and to explain it, we need some more vocabulary concerning them.

We say that, given a graph G and a point p in G, if there is a way of starting at p and traversing a finite number of edges (that is greater than 2) to move through a sequence of distinct vertices of G which eventually ends up at p again, then we call such a path a *cycle* and we call a graph *acyclic* if it doesn't doesn't contain any cycles. *Connected graphs* are what they sound like - they are graphs with the property that for any two of its vertices, there is always a path between them. Further, we say that a connected, acyclic graph G that is also *rooted*, i.e. there is one vertex we call the *root* and every other vertex is considered (and often drawn) "below" that root, is called a *tree*. This makes sense intuitively and Figure 6.2 gives an example of a tree to demonstrate. Similarly, people call the vertices at the ends of branches of trees *leaves* and, if one vertex v_1 is closer to the root than v_2 and their paths to the root overlap, then we call v_1 a *parent* of v_2.

A k-labeled tree is one where each of the tree's vertices are assigned one

---

[8]Some may notice that this definition only applies for integer x (since there is no 3.2th function in our list, for example) - that's because of the caveat that the fast-growing hiearchy only contains functions defined for ordinal inputs.

[9]They aren't actually comparable in size, since $f_{\omega+1}(64) > f_\omega^{64}(6) > G_{64}$.

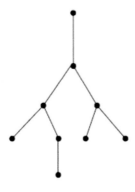

Figure 6.2: A tree.

of k "labels", which for our purposes we may consider as colours. The most complicated definition that Kruskal's Theorem requires us to consider is the idea of a k-labeled tree T_1 being *homeomorphically embeddable* (h.e. from now on) into another k-labeled tree T_2. Given T_1 and T_2, we say that T_1 can be h.e. into T_2 if there is a function $p(x)$ that takes as inputs vertices of T_1 and outputs vertices of T_2 with the properties that

- for every vertex v from T_1, v and $p(v)$ have the same colour;

- for every pair of vertices v_1, v_2 from T_1 and if $xANCy$ denotes the closest common parent of two vertices x and y, then $p(v_1 ANC v_2)$ has the same colour as $p(v_1) ANC p(V_2)$.

To illustrate this somewhat abstract definition, you can look at Figure 6.3

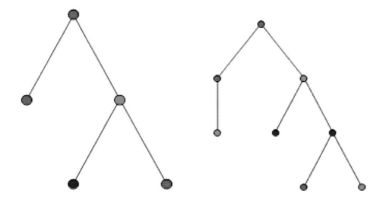

Figure 6.3: The left 3-labeled tree is h.e. into the right 3-labeled tree.

To illustrate this somewhat abstract definition, you can look at Figure 6.3 which gives an example of two k-labeled trees with the first being h.e. into the other. Finally, Kruskal's Tree Theorem states that for any k and for any infinite sequence of k-labelled trees $T_1, T_2, T_3, ...$, where each T_n can have at most n vertices, it is true that for some i and some $j > i$, the tree T_i is h.e. into T_j. While the theorem itself doesn't allow us to describe large numbers, the mathematician Harvey Friedman made the observation that the theorem allows us to define the function $TREE(n)$, which returns the length of the *longest finite sequence* of n-labeled trees such that no T_i is h.e. into a T_j for integers i and $j > i$ and it turns out that k-labellings of trees are amazingly combinatorially rich:

- $TREE(1) = 1$, since T_1 can have at most 1 vertex and at most use 1 colour, so it is simply a single coloured point and that is clearly h.e. into any successive trees of the same colour.

- $TREE(2) = 3$, where the sequence begins with a single vertex of Colour 1, then we have a two-vertex tree of Colour 2 and then we have a single vertex of Colour 2 - can you see why this is the longest such sequence?

- $TREE(3)$ then is so vastly, incredibly large that I struggle to find description for it.

The first thing we can say is that an extremely lower bound that can be proven for it is $f_\omega^{f_\omega(187196)}(2)$, which is clearly immensely bigger than $f_{\omega+1}(64) > G_{64}$! I shall now attempt to explain how far up one needs to go in the fast-growing hierarchy to reach a function that can rival $TREE(n)$.

In the 20th century, the mathematician OSWALD VEBLEN, along with others, was attempting to create a schema for notating and comparing really large infinite ordinals; he first proved that if one has a function $h(x)$ that takes as inputs and outputs ordinals, that is strictly increasing, and whose value for a limit ordinal λ is equivalent to the limit of the sequence $h(a_0), h(a_1), h(a_2), ...$ where the limit of $a_0, a_1, a_2, ...$ is λ,[10] then $h(x)$ has fixed points for some ordinals. This remarkable property of the ordinals is part of what makes them so interesting to study for set-theorists - they seem to be "averse" to notation systems for them. Regardless, Veblen created his *Veblen Hierarchy*, a series of functions $\phi_0(x), \phi_1(x), \phi_2(x), ...$ where $\phi_0(x) = \omega^x$ and each $\phi_{n+1}(x)$ is equal to the xth fixed point of $\phi_n(x)$ - take $\phi_1(1)$, for example, which is the first fixed point of ω^x, i.e. ϵ_0 as discussed previously. Now, the *Feferman-Schütte ordinal* Γ_0 is the smallest ordinal α that satisfies the impressive $\phi_\alpha(0) = \alpha$ or, to paraphrase SOLOMON FEFERMAN himself, Γ_0 is the smallest ordinal that cannot be reached by starting with 0 and repeatedly using addition and the Veblen hierarchy of functions[11] (in fact, it is very difficult to create notation systems that can describe ordinals above Γ_0 as well as all those below it).

What we can now explain is that despite the magnitude of Γ_0, it is possible to show that the growth rate of $TREE(n)$ is *much* greater than that of $f_{\Gamma_0}(x)$.

---

[10]Such a function is commonly called a *normal function* in Set Theory.

[11]Kurt Schütte and Solomon Feferman, *Grundlehren der Mathematischen Wissenschaften*, 1977

But, ultimately, $TREE(n)$ is *piffle* in comparison to what comes next - after all, it grows slow enough for it to still be *computable*...

You recall the distinction between recursive and primitive-recursive functions made by those great minds near the beginning of the 20th century? That result is but one in a bigger theory of computation called, aptly, *Computability Theory*. Computability theorists study fundamentally *what it means to compute something* by examining *models of computation*.

Mathematicians in this era were attempting to go back to the foundations of mathematics and to *formalise* them, that is to make them completely logically rigorous and unambiguous - this was because it had been discovered that without such careful thought and by relying on *implicit* or vague definitions, one can run into paradoxes or questions that don't have an answer, simply because they don't *mean anything*. One of the interesting notions that Church and Turing picked up on this journey was that most people appeared to assume that all functions and questions in mathematics *were answerable* and simply required "proper definitions" and a great amount of thought - their great insight was that the *process* of logically working things out or deducing true statements had mathematical properties in and of itself and thus realised that it was possible to create problems that were well defined (as in, they had a unique answer) but were beyond the capability of any proof-system or computational device to solve; such problems are known as *uncomputable problems*.

Computability Theory is such an amazing area of mathematics that it deserves its own article[12] but we shall simplify here to explain just the concepts required for describing large numbers. Alan Turing came up with the concept of *Turing Machines* (*TMs* from now on), which are theoretical automata that have a *tape* that stretches out infinitely in one direction and is divided up into discrete tape-segments, along with a *tape-head* that is capable of reading and writing symbols on the tape.

It also has a number of *states* that it can be in and a rulebook that determines, given that the TM is in the nth state and that the symbol being read by the tape-head is X, what symbol Y the TM should overwrite X with and whether to then move one tape-segment right or left. It should finally be noted that a TM has two special states denoted YES or NO which, when reached during some stage of a TM-computation, cause the TM to *halt*, i.e. cease all movement. Turing and Church believed that a Turing Machine should be able to perform any algorithm that is well-defined because:

- They (and other mathematicians such as Gödel) had previously created other systems which were supposed to represent computation, such as the λ-*calculus* and the set of general recursive functions, and it turned out that the set of problems they could solve were all equivalent.

---

[12]Perhaps a series of articles!

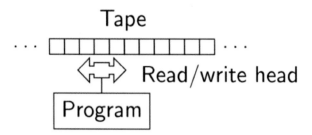

Figure 6.4: An illustration of a Turing Machine.

- They were able to show ways of creating Turing Machines that could evaluate many well known algorithms, such as one that could perform primality-checks or could multiply two numbers together.

- They lived to see JOHN VON NEUMANN, a startlingly brilliant mind, design modern-day computer architectures using their ideas and create the first computers.

In fact, in the 1960s and 70s, many individuals tried to create other deterministic systems that they believed might be able to compute more than Turing Machines but all were eventually proved to be able to solve the same set of problems - thus, they proposed the *Church-Turing Thesis*: that every sequence of logical steps that were performable by a human would also be performable, theoretically, by a Turing Machine, and vice versa; subsequently, any system that was capable of solving the same set of problems as TMs came to be known as *Turing-complete*.

However, as I mentioned previously, those problems Turing-complete systems can solve do not encompass *all* problems, they only contain *computable* problems. But what, then, would an uncomputable problem look like? Well, Alan Turing made the observation that certain Turing Machines are setup such that, after some finite number of steps, they halt in the state *YES* or *NO* and others never halt and simply continue moving along their tape indefinitely. In 1936, he published a paper describing the *Halting Problem*'s uncomputability: it is impossible to create a *general, finite-time algorithm* to decide accurately whether a given TM halts - he showed this through a clever proof-by-contradiction. To illustrate it, we first remember that, by Turing-completeness, any computability theorem that applies to Turing Machines applies to many of the modern-day computer programming languages[13], so we shall precede by writing our proof in the language of *Python*. Let us assume, hoping to reach a contradiction, that there was a program defining a function called *Halts(x)* - this takes in the name of a programmed-function for which you wish to check whether it halts or not and outputs after a finite amount of time either *True* or *False* as per the answer.

---

[13] If we assume they are being run on a computer of unlimited memory and that they are capable of performing genuine arbitrary-precision arithmetic.

Then consider the function:

```
def contradiction(x):
    if Halts(contradiction(x)) == True:
        while True:
            x = x
    else:
        return x
```

For those less familiar with Python, *contradiction(x)* is designed to halt after a finite amount of time if and only if *Halts(x)* says that it will run forever and *contradiction(x)* will loop forever if and only if *Halts(x)* says that it will halt after a finite amount of time. This function is clearly contradictory and so our initial assumption, that there could exist such a function *Halts(x)*, is false. Thus, the Halting Problem is uncomputable!

An interesting observation, then, made by a researcher at **Bell Labs** was that, for a given n, it *must be true* that there is some number k representing the largest number of steps that a *halting* n-state Turing Machine (i.e. a TM which does eventually terminate its computation) will take before stopping - the researcher called these TMs *Busy Beavers* and the corresponding k for each n the nth *Busy Beaver number* ($BB(n)$), because they take a long time to stop and the way that they move up and down their tape during their computation reminded him of a beaver making a dam. So, by definition, there is no n-state Turing Machine that takes longer than $BB(n)$ steps to halt.

However, he pointed out that there cannot exist a general algorithm to compute $B(n)$ for each input n because if there was one, then we would have the following finite-time algorithm to solve the Halting Problem:

- If the TM for which we are trying to determine whether it halts or not has n states, compute $B(n)$.

- Then, start running the TM and wait $B(n)$ steps. If the TM halts by this time, then we know that it halts. If the TM does not halt by this time, then we know, by the definition of $B(n)$, that it does not halt.

but such an algorithm cannot exist by the uncomputability of the Halting Problem and so $B(n)$ is an *uncomputable function*. Further, there cannot be any computable function $g(n)$ which acts as an upper-bound to $B(n)$ for each n, because if that was true then similarly we could wait $g(n)$ steps to determine whether an n-state TM halts (any TM which takes longer than $g(n)$ steps to halt must have already taken more than $BB(n)$ steps, and so cannot halt), giving us again an impossible finite-time algorithm to solve the Halting Problem. So, $B(n)$ isn't just uncomputable - no computable functions can act as an upper-bound to it. So all the functions we have seen previously - $A(m, n), f_\omega(n), f_{\epsilon_0}(n), f_{\Gamma_0}(n), TREE(n)$ - they all must inherently grow slower than $B(n)$ simply by the fact that they are *computable*.

Mathematicians have worked out that $B(0) = 0, B(1) = 1, B(2) = 6, B(3) = 21$ and that $B(4) = 107$ simply by trying and examining all n-state Turing

Machines for $n < 5$ but only lower bounds are known beyond this - for example, we know that $B(7) > 10^{10^{10^{10^{18705353}}}}$ and that $B(5) > 3^{3^{3^{3^{3^{\cdots}}}}}$ where the number of threes is 7625597484987.

But now we can go further *even than this* to functions and numbers so fantastic that I will at the end write down a number that, to my current knowledge, is greater than anything *ever written in any mathematics publication, formal or informal* as it is something of my own design. Because Computability Theory is all about imagining theoretical scenarios related to computation that could never happen–for example, we could never build an *actual* TM, since it requires an infinitely long tape–one thing we can consider is an *oracle*, a purely theoretical construct that would allow us to compute certain uncomputable functions. For example, we could have the oracle $B_0(n)$, which outputs the nth Busy Beaver number, although such a thing could never normally exist in our universe even if it was infinite but remember, this is *pure mathematics* - oracles can exist because we *say so* and they have interesting theoretical properties. For example, if you take the set S_0 of TMs equipped each with $B_0(n)$, the analogous version of the halting problem for S_0 is still uncomputable by those TMs in S_0 and so S_0 has its own associated Busy Beaver function, $B_1(n)$, that grows faster than any function computable by S_0 TMs (and thus grows *far* faster than $B_0(n)$). Similarly, we could then introduce an oracle for $B_1(n)$ and create a new set of Turing Machines S_1 equipped with both the $B_1(n), B_0(n)$ oracles and, once again, S_1 would have its own version of the Halting Problem and thus a well-defined, but uncomputable function $B_2(n)$ that grows even faster than $B_1(n)$.

By iteratively creating more and more sets $S_0, S_1, S_2, ...$, we get faster and faster growing functions assigned to them $B_0(n), B_1(n), B_2(n), ...$ and so, as with the fast-growing hierarchy, we can allow for *ordinal* subscripts where if λ is a limit ordinal, then $B_\lambda(n)$ is the Busy Beaver function for Turing machines with oracles for all ordinals below it. Thus, we can have $B_\omega(n)$ and $B_{\epsilon_0}(n)$ etc. Now, we call an ordinal α *computable* if there is a Turing machine which can tell for any two distinct ordinals β, γ smaller than α whether $\beta > \gamma$ or $\gamma > \beta$ (for example, ω is a computable ordinal, since any good computer can tell, given two nonequal integers, which one is bigger than the other). So, I define the *Ultra-function* $U(n)$ to be $B_{\delta(n)}(n)$, where $\delta(n)$ is the largest ordinal computable for n-state Turing machines - so $U(n)$ grows faster than any individual $B_\alpha(n)$!

Thus, as a final answer to *Who can name the bigger number?*, I advise you to simply write down $K_\Omega = U(TREE(TREE(TREE(3))))$, which is quite concise but which I can verify, thanks to some results on much smaller numbers, is so big that if your opponent attempts to describe another number N, whether the statement $N > K$ is true is in fact *independent* of ZFC-set-theory (i.e. cannot be *proved or disproved* using most modern-day mathematical techniques, regardless of whether it seems obvious).

If you enjoyed this article and would interested in reading more, here are some titles of subjects and numbers that you can look up for further reading:

- Rayo's Number (before I wrote K_Ω's definition above, it *was* the biggest number ever written down in human history)

- Chaitin's constant, the Halting Problem and other uncomputable problems.

- Ordinals and Cardinals in Set Theory.

- Ramsey Theory and Ramsey Numbers.

- Go to the Googology Wiki, a fun website dedicated to listing and describing some of the largest numbers mathematicians know of!

Challenge VI:

1. Determine whether $10^6 \uparrow\uparrow 10^6 > 3 \uparrow\uparrow\uparrow 3$ or $3 \uparrow\uparrow\uparrow 3 > 10^6 \uparrow\uparrow 10^6$.

2. Can you think of a way of describing bigger numbers than using the "uncomputability hierarchy" above along with some form of (minimally-heuristic) argument as to why you believe they are bigger? (NOTE: This would be exceedingly impressive, not just to us but to the entire mathematics community.)

Chapter 7

ARM in the Big Apple

Salutations!

Adventures in Recreational Mathematics normally exposes some particular set of interesting ideas, but this issue is special. The author over the summer holidays had to be *regrettably* pulled away from his computers, mathematics papers and dark room to New York. The author, of course, travelled there at personal expense for purely research reasons - to visit the National Museum of Mathematics or MoMath, as it's called on posters and on the entrance[1]. What follows is a discussion of the museum, some of the exhibits there and the mathematics behind them, though I should be clear that the museum had much more to offer than what I mention here and so if you're ever in New York & are mathematically curious, then give it a visit!

The Goudreau Museum, named after an American mathematics teacher BERNHARD GOUDREAU, ran on Long Island from 1980 to 2006 and was the only institution in the U.S. in which public exhibits were put on purely for education or entertainment about mathematics.

After its closure, numerous mathematics communicators, academics including John Overdeck[2] formed a group hoping to create a replacement. Their work came to fruition when this museum was opened in 2012, accompanied by a variety of amusing headlines from local newspapers such as DNA-Info's "New Math Museum hopes to Make Numbers Fun for Kids" along with a brilliant "rendering of what the museum might look like"[3], found in Figure 7.1.

Figure 7.1: Architect-standards of detail.

The first exhibit is the slide-like apparatus, shown in Figure 7.2, for young

---

[1] Any visits to other restaurants, museums or skyscrapers during this *entirely academic* field-trip were, naturally, complete accidents.

[2] Founder of the quantitative hedge-fund Two Sigma Investments LP. Interestingly, the recently refurbished mathematics gallery in London's Science Museum was sponsored by Winton Capital, a British quantitative-trading firm.

[3] Not only is the image highly detailed, but the article's headline is extremely accurate in describing what mathematicians do (which is, of course, to sit around, look at numbers and read copies of this book).

children.

Figure 7.2: An uncomfortable ride?

As you can see, the children get into a perspex carriage on the left and pull themselves forwards and backwards using the two black ropes, moving along by rolling over the various oddly shaped objects in the foreground. So, surely the children will have quite a bumpy ride, rolling over those pointy objects? No, these are interesting examples of *objects of constant width* that are not spheres. To see that such things exist, let's look in 2-space. A circle is a shape of constant width, but we can make another by taking the vertices of an equilateral triangle and, for each point, drawing a circular-arc between the opposite two points. This, pictured below, is called a Reuleaux triangle and can be easily drawn on a piece of paper using a compass and it should be clear why this is a shape of constant width[4].

We can do the same for any odd n-gon, simply by drawing an arc for each point between the pair of points opposite. If we take any of these and rotate them around one of their lines of symmetry, then we get a *Meissner body* — a surface of constant width. See if you can figure out which Reuleaux n-gon corresponds to which Meissner body back in Figure 7.2!

Another exhibit near the entrance to the gallery is a tall column of elastic fibers with a chair in the middle which, if swiveled, causes the fibers to twist inwards, as shown in Figure 7.4.

This display constructs a hyperboloid (the surface of revolution of a hyperbola) using just straight cords of material - have you ever considered how many other surfaces can be constructed in this way? Such structures are called *ruled surfaces* and are prime objects of study in the field of projective geometry. They can be naturally defined as those surfaces which can be constructed by tracing a

---

[4]It's what one might call a piecewise-circular shape!

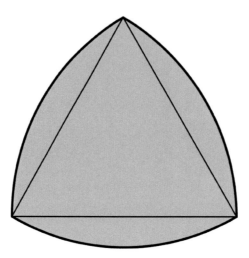

Figure 7.3: A Reuleaux triangle.

vector through 3-space and a famous fact about them is that any surface with *zero gaussian curvature*[5], i.e. any that can be unrolled and flattened to make a plane (such as cylinders, but *not* spheres), is a ruled surface. However, more special are the *doubly ruled* surfaces, in which each point is part of 2 distinct straight lines that lie on the surface; there are only 3 types of d.r. surface: the plane, the hyperbolic paraboloid and the single-sheeted hyperboloid (all pictured below).

The red and the yellow fibers in the exhibit show the two different rulings of the single-sheet hyperboloid!

One of the most famous displays in the National Museum of Mathematics is a bumpy, circular course that children ride around using specially designed tricycles with *square* wheels.

The curves on the ground are designed such that one gets a completely smooth ride, despite using such bizarre wheels. The natural question to ask when one sees the exhibit is: what possible curve could the bumps be such that this interesting behaviour is observed? Well, it turns out to be the same curve one gets when a length of rope hangs between two poles on a fence, just upside-down!

The curve is called the *catenary* and it is not, perhaps surprisingly, a parabola but instead the graph of the hyperbolic cosine function, $cosh(x)$. The normal trigonometric functions $cos(\theta)$ and $sin(\theta)$ are often defined as the x and y coordinates of the intersection between the unit circle and a ray from the origin at an angle θ from the x-axis; in the same sense, the hyperbolic trigonometric functions $cosh(\theta)$ and $sinh(\theta)$ are defined as the x and y coordinates of the intersection between the unit hyperbola $x^2 - y^2 = 1$ and a ray from the origin at an angle θ from the x-axis. While at first sight, this may seem like a bizarre

---

[5]To learn about Gaussian curvature, one can read Chapter 6 of this book.

Figure 7.4: A doubly-ruled surface.

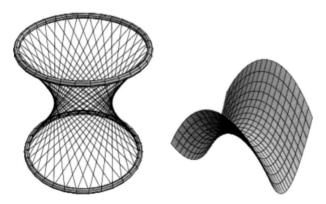

Figure 7.5: On the left, the hyperbolic paraboloid; on the right, the single-sheeted hyperboloid.

Figure 7.6: Tricycles with square wheels!

Figure 7.7: A catenary.

idea used only in truly strange situations, the functions appear commonly in the physics of heat transfer and do actually describe trigonometry for certain right-triangles (it's just that these triangles exist only in the exotic, negatively-curved world of hyperbolic geometry).

It's said that one of the members of the Bernoulli family once posed the problem of finding the curve that describes the road on which a square wheel can roll at constant height to the European mathematicians of the era. He took 3 months over his solution, his brother took 5 months over the solution but Newton, ever the mathematical god, solved it **the night he received the problem** (supposedly considering the Bernoullis to be extremely arrogant) and sent his proof via an anonymous letter. Despite his lack of a signature, it is further said that all who read it could immediately guess the author...

Catenaries (i.e. functions of the form $a \times cosh(bx + c) + d$) have many surprising properties[6] — to see another, consider the surface-of-revolution around the x-axis of a catenary[7], called a *catenoid*. If you have 2 points above the x-axis in the Cartesian plane and you want to choose the curve C to draw between them so that you minimise the surface area of the s-of-revolution one would form by rotation around the x-axis, the catenary between the two points is the optimal choice, so the catenoid is the "minimal surface"! An amazing way of illustrating this is with bubbles: the surface constantly attempts to minimise its surface area because of surface tension — this is why, for example, bubbles on their own form spheres, as spheres are the shape that maximise the volume-to-surface-area ratio[8]. If we form a bubble between two rings, what forms is none other than the catenoid, as shown in Figure 7.8!

While I was there, I got to see the special exhibition on mirrors and reflections, full of a variety of optical wonders and great mathematics. On the one hand, mirrors are a great way of demonstrating the idea of functions or maps on the plane. The image below shows two ships crashing together in the midst of a great storm to the ignorant viewer but becomes the portrait of a well-dressed man with a significant beard when one places down a cylindrical mirror at a precise point on the canvas — the mirror acts as a map or *bijection* between the portrait and the landscape in that for every point in the former, there is a single corresponding point in the latter via reflection. It is just the fact that this function is relatively simple to compute and understand that enabled the artist to create this artwork.

They also had a cleverly designed conical cavity in a table (pictured below), with its four sides each being a plane mirror. This meant that one could place various shaped pieces of wood in it carefully and naturally create polyhedra - in the image, I have created a *cuboctahedron*.

While this may seem fairly innocuous, it hints at a major aspect of modern mathematics: abstract algebra, or more specifically, group theory. The key idea

---

[6] For example, I have never actually managed to find a proof of the fact that chains, ropes etc. hang in the shape of a catenary, which is quite a remarkable fact.

[7] This is the surface the function "carves out" in space if you were to imagine spinning the curve around the axis - there are many diagrams of this online.

[8] This can be proved using the isoperimetric inequality in 3 dimensions.

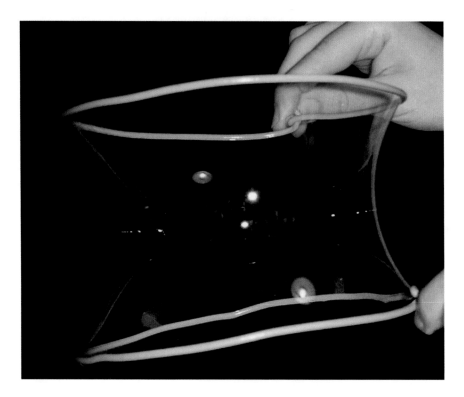

Figure 7.8: The bubble formed between two rings or a surface of revolution?

Figure 7.9: An optical witticism.

Figure 7.10: A rhombicuboctahedron formed with mirrors.

is that one could see, as in the portrait example, the various mirrors not as physical objects but rather as abstract geometric transformations which one can perform on the points of the chunk of wood I have used. Therefore, imagine chopping up our rhombicuboctahedron by placing mirror-planes through each of the faces as you see in the image and giving the process of reflection through each mirror a name as a function. Then, we have a collection of functions with the following interesting properties:

- If you compose any two functions in our set, your result is some other function already in our set (i.e. the set has *closure* under composition).

- The set has a function which, if you compose it with any function f returns f (we'll call this the *identity* and it equates to "doing nothing" with a point).

- For any function in the set f, there is another function g such that $g(f(x))$ is the identity (we call this the inverse element and in this case the inverse of f is just f, as reflecting something through a mirror twice gets you back to the same point).

Those four properties are known as the *group axioms* and, unsurprisingly, any set which has an associated way of combining two elements to get a third which satisfies the above axioms is called a *group*. There is another, final, axiom, known as *associativity*, that I refrained from mentioning because it's quite difficult to explain the above context: if a, b, c are elements of our group, $(a*b)*c = a*(b*c)$ (it should be clear why this is for the functions above, as there's no distinction between performing a function f before g & h and f & g before h). It's an amazing fact that the symmetries of most geometric objects,

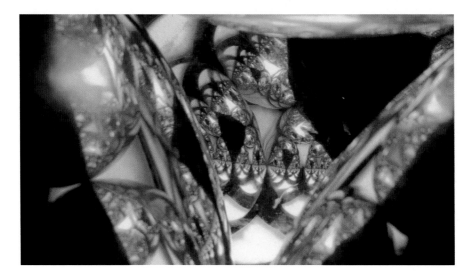

Figure 7.11: A complex lightshow.

like polygons or polyhedra or circles etc., form groups and groups appear in
many other places. If you think about the integers and read the above axioms
replacing the word "function" with "integer" each time it appears, you'll soon
realise that they form a group under addition (0 here is the identity element as
$n + 0 = 0 + n = n$), and that in the same way, the real numbers bar zero form
a group under multiplication. Group Theory creates theorems about groups in
the complete, general abstract and then mathematicians can use these theorems
to investigate properties of a great variety of situations, from combinatorics to
number theory! In later ARMs, I hope that we will investigate some of these
amazing applications in more detail than we have time for here.

The exhibition also showed a remarkable example of chaos in a relatively
simple geometric situation. I shall first show the image — try and see if you
can figure out what's going on before reading on!

Well, imagine taking a tetrahedron and centering at each vertex a sphere
of equal size, such that the spheres just touch. Then, imagine assigning an
orange, green and pink light each a different face, shining inwards, and finally
placing a camera looking inwards on the fourth face — this is what the image
shows. Shining light inbetween reflective spheres turns out to be a *chaotic
system*, which means that two initial states of the system (in this case, two
incident beams of light) which are very close to each other will quickly move
far apart ("sensitivity to initial conditions") and any interval of inputs, i.e.
all incident beams between 2 and 4 degrees to their first sphere's surface, will
quickly spread out to touch almost all pathways and points between and on the
spheres ("topological mixing"). Dynamical systems - those with a collection of
parameters that define their state and which come with a transition function
that gives you the $(n + 1)$-th state if you give it the nth state of the system

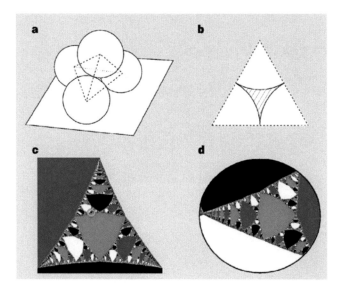

Figure 7.12: The diagram found in the *Nature* 1999 article.

- with these properties are the prime objects of study in Chaos Theory, which attempts to classify and understand them[9].

The fact that four spheres arranged like this form a chaotic light-scattering system was first described in a Nature paper of 1999, from which we get the following explanatory diagram (**Figure 12**) and which attempted to conduct research into the greatly complex light display you see. They noted how the different colours of light seem to form dense and complicated triangular patterns, which they called *Wada Basins*, and showed that the W. Basins created by light sources from two different faces of the tetrahedron were, interestingly, completely disjoint. They said that even if you "fired" a beam into the arrangement from the face F at an odd angle, you would find that your light beam's path would very quickly "attract" into the face's associated Wada Basin and that, further, many light rays would fall into completely periodic orbits in which they were doomed to forever bounce at regular intervals between the four spheres... This doesn't quite happen in real life, of course, as the spheres are not completely reflective and so will eventually absorb the light, but the display is nevertheless both beautiful and confusing.

If you, the reader, can find an intuitive reason as to why the Wada Basins of different light sources are completely disjoint in this setup, please contact me, as I still find this utterly baffling.

For the final exhibit, imagine a camera connected up to a projector, so that the projector shows live what the camera sees, and then imagine pointing the camera at the projector's screen. What would you see? You would see an image of the projector's screen, showing an image of the projector's screen, showing

---

[9]These also are part of future plans for ARM!

Figure 7.13: Friends old...

Figure 7.14: ... and new.

an image of the projector's screen, showing an... This would look like an infinite series of rectangles, each one containing the last. If you wish to try this yourself, just connect a camera up to your computer and point it at the computer screen or, more simply, take two smartphones, make them face each other and put each of them on their camera apps, so that they use their "selfie" or forward-facing sensor.

Now imagine having three of these, so that the projectors all slightly overlap and the cameras can each see all of the projections. This was the setup on the bottom floor of MoMath and it allowed you to create a great variety of impressive and complicated images. Here is an image taken of one configuration, shown to me by one of the staff at the museum:

It's a Sierpinski triangle! Remarkable! By rotating each camera smoothly, you can continuously form new fractals in real time and at a resolution bounded only by the resolution of the cameras. Here's another such fractal - Figure 7.14.

There's little to say on this particular display without going into great detail, but we will speak more about fractals in coming issues of *The Librarian*

Supplement. Instead, I shall end with this issue's two challenges, which break form: normally, we have the first challenge be completely simple to anyone who's read the article properly and the second challenge be extraordinarily difficult, so that perhaps even professional mathematicians might take a couple of hours over it[10]. Both parts to the challenge in this issue are non-obvious, though neither should be difficult in the extreme.

1. When shown the stairwell to the lower floor, I noticed a large paraboloid hanging in the middle of the spiral with numerous ropes within it. When asking about it, one of the staff explained that it was a calculator. If you wish to multiply 2 numbers a and b, you go up to the height z_1 such that the circular intersection between the paraboloid and the locus of points that satisfy $z = z_1$ has a radius of a and then find the height z_2 such that the same is true for b and then draw a line from one point on the paraboloid at height z_1 to a point on the opposite side of the z-axis at height z_2 - the intersection between the z-axis and this line has z-coordinate $a \times b$! If that was a bit of a mouthful, simply draw a diagram and think through again what I was saying. Now, when asking about this, the member of staff I asked explained that proving this fact was part of the interview for his job and so, after thinking about it for a few minutes, I proved it myself. Can you?

2. Seeing the Sierpinski triangle appear in the camera-setup was surprising but can you figure out what the orientations of the cameras must be to do this? As an extension (difficult and open-ended), can you think of a way of characterising precisely those fractals which can be formed using the setup, such that if someone "shows you" a fractal, you can tell them whether the cameras can make it? How about with 4 cameras?

---

[10]The idea is not necessarily for one to solve it (though this would be wonderful!) but instead to give the reader something to sink their teeth into and think about.

Chapter 8

Modern Combinatorics

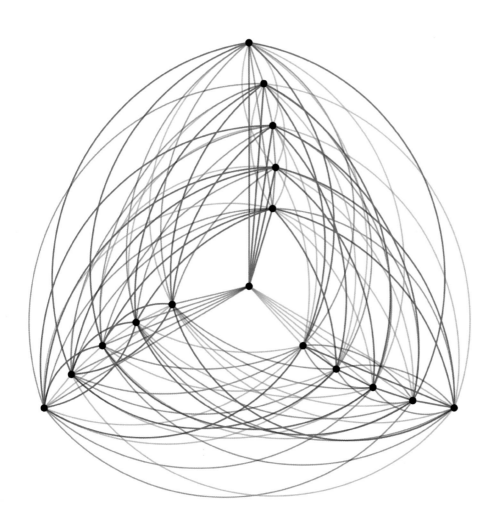

Salutations!

We're back to our normal form after last issue's *ARM in the Big Apple* with an issue this time on modern-day combinatorics. We thought this would be topical given the recent announcement of a proof of the *g*-conjecture, a major problem in the combinatorics of polytopes!

What *is* combinatorics? Many people first meet it in secondary education as a subject in which one *enumerates* objects of various kinds or the number of ways of doing certain structural operations on a set, such as the number of distinct integers that be formed from the set of digits $\{1, 3, 5, 0\}$ without repetition or the number of ways of dividing a group of n children into two smaller groups. However, since the 19th century, we have developed a variety of remarkable tools that allow us to solve all problems you'll encounter in school relatively easily and so most combinatoricists today more generally study any structure of discrete objects[1] that doesn't fall under the purview of such subjects as Number Theory.

To illustrate what I mean, there's no better way than to give an example: *the Pigeonhole Principle.* This apparently innocuous statement is that if you have n holes to place pigeons in but you have $(n+1)$ pigeons, then two pigeons must go in the same hole - do not let its simplicity beguile you, for it was shown early on that it's an extremely powerful tool. Consider the following problem:

Prove that in any set S of 27 odd integers smaller than 100, there is a pair of integers whose sum is 102.

This is proved relatively simply via the principle. Consider all the odd numbers from 1 to 100 and let's try to put them into pairs that add to 102, e.g. $(3, 99), (5, 97), (7, 95), ...,$ and we're left with the integers 51 and 1 which have no pairs. There are, according to this arrangement, 22 pairs along with the individual 51 and 1 - let's see these 24 groupings as the "holes" - and so with 27 odd numbers in S we know that there must 2 numbers from our set in one of the pairs, thus two numbers whose sum is 102.

There are many similar problems that are apparently extremely complex to those unfamiliar with the principle but which have simple solutions by trying to find the "holes" and the "pigeons" in the problem. However, there is a *much more powerful* generalisation of it created by FRANK P. RAMSEY in 1928[2], which started a whole new field of mathematics - *Ramsey Theory.*

To explain Ramsey's Theorem, we need to first remind ourselves of the definition of a graph. A graph is simply a collection of points on the plane, known as the *vertices* of the graph, and a collection of pairings of those points, known as the *edges* of the graph and often drawn as curves going between each such pair. Examples of graphs are shown in Figure 1, which those of you who are more familiar with Graph Theory will recognise. A *subgraph* of a graph G is simply a graph whose vertices are a subset of G's vertices and whose edges

---

[1]i.e. they won't be looking at continuous things like functions on $\mathbb{R}$ or the geometry of smooth surfaces

[2]It may interest the reader to know that Ramsey's Theorem was not written in a text on combinatorics or much related but instead stated and proved as a minor result in a monograph on the foundations of mathematical logic.

are a subset of G's edges.

An *n-edge-colouring* of a graph G is simply a particular assignment of one of n colours to each of the edges of G (note that you don't need to use *all* of the n colours in an n-edge-colouring), e.g. we could colour each of the edges of the graphs either *red* or *blue* to give them 2-edge-colourings. You also need to know about the *complete graph on n vertices*, denoted K_n, which is the graph with n vertices that has an edge between every two vertices, e.g. K_3 is a triangle and K_4 is a square with the diagonals also drawn in.

Ramsey's Theorem is *so general* that we shall begin by stating and understanding some special cases before looking at the full result; it essentially says that for large enough n, an edge-coloured K_n becomes so interconnected that you can't help but have complete subgraphs where all the edges are the same colour. For example, you might wonder if there's an n such that K_n, when given a 2-edge-colouring, is *forced* to have a monochromatic subgraph K_3, i.e. a triangle where all the edges are the same colour. This n is written in Ramsey-notation as $R(3,3)$ where:

- R stands for *Ramsey* and is a function.

- The fact that it has 2 arguments represents the fact that the graph in question will be 2-edge-coloured.

- The "3"s in the arguments refer to the fact that we are looking for a subgraph K_3 with all the edges the same colour.

We happen to know that $R(3,3) = 6$, i.e. that for *any* 2-edge-colouring of $K_{R(3,3)} = K_6$ there *must* be either a subgraph K_3 with all the edges being the first type of colour or there must be a subgraph K_3 with all the edges being the second type of colour.

A more general set of cases that Ramsey's theorem discusses is the statement that for any q colours, there is a number, denoted $R(3, 3, ..., 3)$ (where there are q "3"s), such that any q-edge-coloured $K_{R(3,3,...,3)}$ contains as a subgraph a monochromatic triangle.

An even more general statement of Ramsey's Theorem says that for any q colours and any n, there is a number, denoted $R(n, n, ..., n)$ (where, again, there are q "n"s), such that any q-edge-coloured $K_{R(n,n,...,n)}$ contains as a subgraph a monochromatic K_n.

To illustrate each of these, consider how previously we discussed that $R(3, 3)$ is the size of complete graph which is forced to contain a monochromatic triangle when 2-edge-coloured, so $R(3, 3, 3$ is the size of complete graph which is forced to contain a monochromatic triangle when 3-edge-coloured and $R(10, 10, 10)$ is the size of complete graph that is forced to contain a monochromatic K_{10} subgraph when 3-edge-coloured. Can you understand what $R(20, 20, 20, 20)$ represents?

The full Ramsey's Theorem says that for any q and finite series of numbers $(n_0, n_1, n_2, ..., n_q)$ there is a number, denoted $R(n_0, n_1, n_2, ..., n_q)$, such that any q-edge-coloured $K_{R(n_0,n_1,n_2,...,n_q)}$ necessarily contains either a monochromatic subgraph K_{n_0} or a monochromatic K_{n_1} or .... or a monochromatic K_{n_q}.

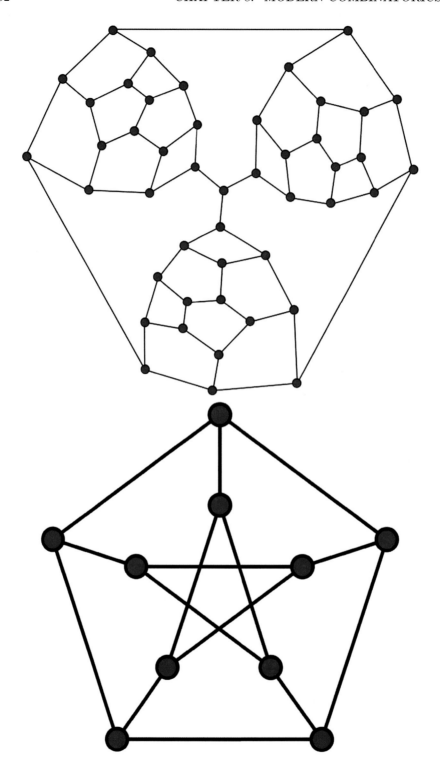

Figure 8.1: The *Tutte* and *Petersen* graphs.

For those who have not seen the result before, I suggest strongly that you re-read and think about the previous few paragraphs carefully, as Ramsey's Theorem is quite a technical result and one whose proof we could not possibly give here.

Why on earth would such a result be useful? An example is the proof of *Schur's Theorem*, which says that if you divide the natural numbers $\mathbb{N}$ into a finite number of partitions then there will necessarily be one partition which contains three numbers x, y, z such that $x + y = z$. If you think about it for a bit, you'll realise that this result is really quite surprising!

The proof first notes that we could see the process of putting each number into one of the q finite partitions as assigning each number one of q colours. Then, suppose we have the set S of all the integers from 1 to $R(3, ..., 3)$ (where there are q "3"s) and they have been each given one of the q colours. Now, consider the complete graph $K_{R(3,...,3)}$, where each vertex represents one of the numbers from 1 to $R(3, ..., 3)$ and let's colour the edge going between the integers a and b by the colour that the number $|a - b|$ has in S.

That setup makes it sound like I'm about to do something extremely complicated but instead just remember the definition of $R(3, ..., 3)$. We know that any q-edge-coloured $K_{R(3,...,3)}$ definitionally contains a monochromatic triangle. So, in our situation that means that there is a monochromatic triangle of 3 integers a, b, c, whose edges represent the numbers $x = |a - b|$, $y = |b - c|$ and $z = |a - c|$. The monochromicity just means that x, y, z are in the same partition and their definition leads to the fact that $x + y = z$.

If you re-read this proof (which remarkably only requires two wordy paragraphs), you'll that most of "the work" is simply done by the existence/definition of Ramsey numbers. Perhaps through this you'll appreciate the great power of Ramsey's Theorem and see if you can prove anything yourself with it! Ramsey Theory is the study of the results one can deduce from Ramsey's Theorem and similar "structural" results, which make statements about the conditions under which *ordered objects* are forced to appear. For example, here the ordered object was the existence of a K_n with sufficiently large complete graphs but the content of the *Hales-Jewett Theorem* is that for any number of players p and number n, there is a dimension d such that in all games of p-player noughts-and-crosses somebody *must win* no matter how the game plays out.

Another massive subject in the study of discrete structures is *arithmetic combinatorics*, from which we obtain two of the author's favourite theorems of all time. This field began with an interesting question by PAUL ERDOS and PAL TURAN, some of the 20th century's most influential combintoricists, relating to the *natural density* of a subset S of $\mathbb{N}$. The natural density of S is defined as

$$\lim_{n \to \infty} \frac{|S \cap \{1, 2, 3, ..., n\}|}{n}$$

which can be read as the proportion of the numbers from 1 to n in your set S as $n \to \infty$. So, any finite set has natural density 0 but there are also very many infinite sets with 0 natural density, such as the square numbers or the primes. The fact about square numbers follows from the fact that the number of square numbers below n is $\sqrt{n}$ rounded down to the nearest integer and that $\lim_{n\to\infty} \frac{\sqrt{n}}{n} = 0$ and the fact about the primes the interested reader will perhaps research into. Another example, just to make the idea of density intuitively clear in the reader's mind, is that the natural density of the even numbers is 0.5.

It's clear that *arithmetic progressions*, i.e. sequences of integers of the form $a + b, a + 2b, a + 3b, ...$, are abundant in $\mathbb{N}$ but Erdos and Turan wondered how "big" a subset S of the naturals would have to be to force an infinite number of arithmetic progressions of length $k > 2$ and their way of quantifying precisely what they meant by "big" was to use the definition of natural density above, i.e. they conjectured that any set with positive natural density would have infinitely many arithmetic progressions of length k for all positive integers k.

In 1952, the remarkable mathematician KLAUS ROTH announced a proof of the above conjecture for the case where $k = 3$ in a paper entitled **On Certain Sets of Integers**[3], one of the 2 results that were cited in his receiving of the Fields Medal in 1958. Following this, not much work was completed for some years and many tried to understand the deep reasons behind the fact that Roth's method to prove the $k = 3$ failed to work on any other cases. The conjecture was eventually given a proved positive solution in 1975 by ENDRE SZEMEREDI and it is now commonly known as *Szemeredi's Theorem*. The proof is legendarily one of the hardest to understand in all of mathematics, with a large diagram required to describe the complex non-linear collection of theorems, lemmas and proofs that lead to the final QED - this is shown in Figure 8.2 as taken from the original paper by Szemeredi himself.

It had been known since 1837 that if a and b are integers that share no common factors other than 1 then the infinite arithmetic progression $a + b, a + 2b, a + 3b, ...$ contains infinitely many primes, a most celebrated result by the founder of analytic number theory, PETER LEJEUNE DIRICHLET. However, in the early 2000s, mathematicians BEN GREEN and TERENCE TAO were examining Szemeredi's proof-method to see if they could prove the reverse-analogue of Dirichlet's result: that the primes contain arbitrarily long arithmetic progressions. This did not follow immediately from Szemeredi's Theorem as we recall that the primes are a set of natural density 0; instead what they did was to create what they termed a *transference principle* that allowed them to prove an analogue of Szemeredi's Theorem for subsets of the naturals that are *pseudorandom* in a specific sense and then they showed that the primes form a dense subset of an explicit example of such a pseudorandom set, allowing them to prove their desired result. This technique has since become one of the standard tools of arithmetic combinatorics (they term the transference principle result a *relative Szemeredi Theorem*) and the so-called *Green-Tao Theorem* itself was one of the many many remarkable results cited at the 2006 ceremony at which Terence Tao received his Fields Medal. Having received it at age 31[4], this makes him one of the youngest recipients of the Fields Medal ever and one of the greatest

---

[3]This title, I believe, is one of mathematics' great masterpieces of understatement.

[4]Incidentally, this is the age of PETER SCHOLZE, one of this year's Fields Medallists for his work in arithmetic geometry.

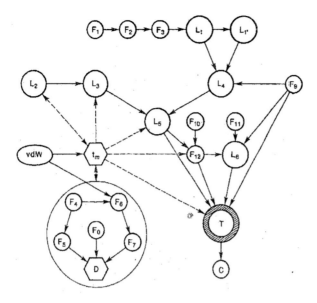

Figure 8.2: A complex diagram: each L_i refers to a specific lemma he proves, each F_i refers to a specific fact he mentions/proves, t_m is the definition of the mth term of a specific sequence he defines, D is an important series of definitions he makes and vdW refers to *van der Waerden's Theorem*.

mathematicians of our generation.

We now turn our view to the study of combinatorics on polytopes, which is extremely relevant given the publishing of a paper that solved the famous g-conjecture on Christmas day of 2018!

What is a polytope? 0-dimensional polytopes are points, 1-dimensional polytopes are line-segments, 2-dimensional polytopes are polygons and 3-dimensional polytopes are polyhedra. The key common feature of these examples is that each of them is a region of n-space of finite volume that is bounded by the n-dimensional equivalent of planes[5] (i.e. in 2D, these are lines, 3D these are planes as we know them and in 4D they are 3-dimensional *subspaces*) - so a *polytope* generalises the idea of polygons and polyhedra to n dimensions. In the same way that polygons have line-segments and polyhedra have polygons as the "faces" on their boundaries, it is also true that n-topes have $(n-1)$-topes as their "faces" - in fact the n-dimensional version of "faces" are often referred to as *n-facets* and the ordered list $(f_0, f_1, f_2, ...)$, where f_i is the number of i-facets the polytope has, is called its *f-vector*, so a square-based pyramid has five 2-facets (the faces), eight 1-facets (the edges) and five 0-facets (the vertices) and so has f-vector $(5, 8, 5)$. A *convex set* of points in Euclidean space is one

---

[5]For those who know a little linear algebra, a better and more precise definition of an n-plane would be the kind of set one produces as the span of n linearly-independent vectors in Euclidean space.

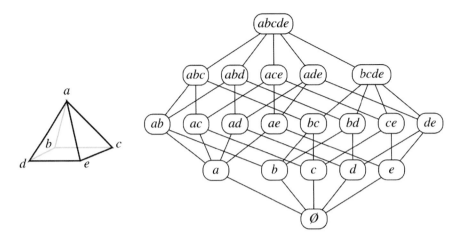

Figure 8.3: The face lattice of a square-based pyramid, with edges & faces labeled according to the vertices.

with the property that for any two points within the set's boundaries, the line segment connecting those two points is contained entirely within the set also - combinatoricists study *convex polytopes,* i.e. those with the property that the set of points within their boundaries is a convex set.

The brilliance of convex polytopes as combinatorial objects of study is their interplay between geometric and purely combinatorial properties. Consider the fact that on a polygon, each side can be considered to be "included" or "within" the whole shape and similarly each vertex of the polygon can be seen as "within" the two line-segments it is part of. Extending these observations, we may note that every facet of a polytope contains and is contained by a collection of other facets of the polytope and so it is possible to construct a hierarchy, known as the polytope's *face lattice*[6], for the facets showing which ones include which (it is a tradition for reasons of simplicity to include the entire polytope and also the empty set ∅ in this hierarchy) - the face lattice of our square-based pyramid is shown in Figure 3.

To gain insight into the structural nature of convex polytopes, then, we could prove results about face-lattices which have no mention of geometric properties whatsoever; indeed, one of the fundamental results of this field is that polytopes can be defined entirely via combinatorial properties of their face-lattices (i.e. there is a short list of combinatorial constraints a lattice has to satisfy that means it must be the face-lattice of some polytope.

Another way of abstracting the discrete nature of a polytope is to simply take the graph whose vertices are those of the polytope and which has an edge between two vertices if the corresponding vertices of the polytope have an edge

---

[6]Those who read the ARM article *Cantor's Attic* may recognise this to be a poset, whose order-relation is "included within".

between them. One of the famous results in understanding these graphs is *Steinitz's Theorem*. He noted that for any polyhedra:

- Their graphs are *planar*, i.e. can be drawn on the plane without the lines representing the graphs' edges crossing over one another.

- Their graphs are *simple*, i.e. they have the property that between any two vertices there is at most one edge that goes between them and there are no edges from vertices to themselves.

- Their graphs are 3-*connected*, i.e. you can remove any 3 vertices from them and the graph you end up with still has the property that between any two points there is a path along the edges of the graph between them.

Can you prove/see why the first and third statements are definitely true?

Steinitz's Theorem, then, says that the converse is true, i.e. that *any graph* that satisfies above 3 criteria is the vertex-edge graph of a polyhedron. This elegant result allows us, amongst other things, to enumerate all the polyhedra that have n vertices relatively simply via computer searches, the first few terms of which are $0, 0, 0, 1, 2, 7, 34, 257, 2606, 32300, 440564, 6384634, ...$, i.e. there are 0 polyhedra with either 1,2 or 3 vertices, 1 with 4 (the tetrahedron) and 32300 with 10. The sequence, for those who wish to examine it in more detail, is *A000944* in the *Online Encyclopaedia of Integer Sequences* and shows impressively fast growth!

The first bricks on the road to the g-conjecture were paved by Swiss genius LEONARD EULER, when he noticed the now-famous relation between the number V of vertices, the number E of edges and the number F of faces of any polyhedron: $V - E + F = 2$. If you've never seen this before, I implore you to try it out on a number of well-known polyhedra - by knowing certain facts, such as that every point on an icosahedron is connected by an edge to 5 others, and using *Euler's polyhedral formula* you can work out the number of faces, edges and vertices of all the platonic solids on paper without looking up anything.

There are numerous proofs of this fact, none of which are particularly long or complicated, and so we instruct the interested reader to look these up. Many of these proofs rely on properties of *triangulations* of polyhedra, i.e. ways of dissecting their faces such that the polyhedron only has triangular faces, and, due to the great abundance of results about triangulations in combinatorics, various individuals wondered if analogues of Euler's result applied in lower or higher dimensional spaces with corresponding lower or higher dimensional analogues of triangulated-spheres.

Let's examine the alternating sum of the entries in the f-vectors of lower-dimensional triangulated-spheres (i.e. $f_0 - f_1 + f_2 - ...$), starting with the 2D plane. A circle is the 2D analogue of a sphere and so the "triangulations" of a circle are the polygons (here, the line-segments are the 1D analogue of triangles), which obey the relation that $V - E = 0$. In the 1D Euclidean space (the line) we see that "spheres", being just the set of all points in n-space that are equidistant from a special point called its center, are just 2 points either side of the center.

So, the relation obeyed here is just $V = 2$, as 0D triangles are just points. So, the alternating sums, starting with dimension 1, go 2, 0, 2, ... How does it continue? You might guess that it alternates as $1 - (-1)^d$ and you'd be right, but to understand this we need to quick make clear what we mean by "higher dimensional triangulations of spheres".

An *n-simplex* (plural *simplices*) generalises the notion of a triangle - it's an n-dimensional polytope which is the *convex hull* of its $(n + 1)$ vertices, i.e. the smallest (in terms of volume) convex set that contains its vertices. A *simplicial complex C* is a set of simplices of various dimensions with the property that any facet of a simplex in C is also a member of C and the intersection of any two simplices is a facet of both and is therefore also in C. Therefore, a *simplicial d-sphere* is a simplicial complex which is "topologically equivalent" to the d-dimensional sphere; precisely what this topological equivalence *means* is a little too technical for this article but one can acceptably describe two objects to be "topologically equivalent" (formally, *homeomorphic*) if one can continuously deform the first object into the second object simply by treating it as a rubber-sheet object of infinite elasticity (i.e. one can *stretch* but not introduce holes or rips) - I find simplicial spheres lovely objects to imagine (dazzlingly complex collections of line-segments, triangles and more, linking together to make a sphere).

Now that we've established the higher-dimensional equivalents of triangulations of spheres, we can return to our discussion of Euler's formula. It turns out that it's one part of a much bigger series of ideas which are most simply explained in terms of the h-vector of a simplicial, which can most easily be calculated by *Stanley's Trick*, where one writes out the f-vector of the complex as the side of a triangle and you make the other side 1s and then you fill the triangle out by filling each entry with the absolute difference of the two entries above it - examples of this construction are shown for various simplicial spheres[7] in Figure 4.

The bottom row of each filled-out triangle is the h-vector of the corresponding complex - notice that the furthest right entry is 1. If you work through the series of subtractions going down the triangle (try it now!) in the octahedron you'll see that this is the result of $|||6 - 1| - 12| - 8| = |||1 - f_0| - f_1| - f_2| = -1 + f_0 - f_1 - f_2 = 1$, i.e. the alternating sum of the f-vector.

However, notice that one of the simplicial spheres in Figure 4 is a 4D polytope (the 16-*cell*) and its far entry naturally comes out to be one (the alternating sum of the f-vector $+1$). So, the h-vector appears initially to be a really arbitrary thing to define but with it, the dependency on the parity of the dimensions that the "Euler formula" is gone, instead we can say that the far-right entry of the h-vector of any simplicial sphere is 1. This is the first of a collection of generalisations of Euler's formula discovered and proved in the 1920s called the *Dehn-Sommerville relations* - the others prove that, as we see in Figure 5, the h-vector of any simplicial sphere is *palindromic*.

Finally, we can define the g-vector of a simplicial sphere to be the differences

---

[7]The *16-cell* is a 4D analogue of the octahedron.

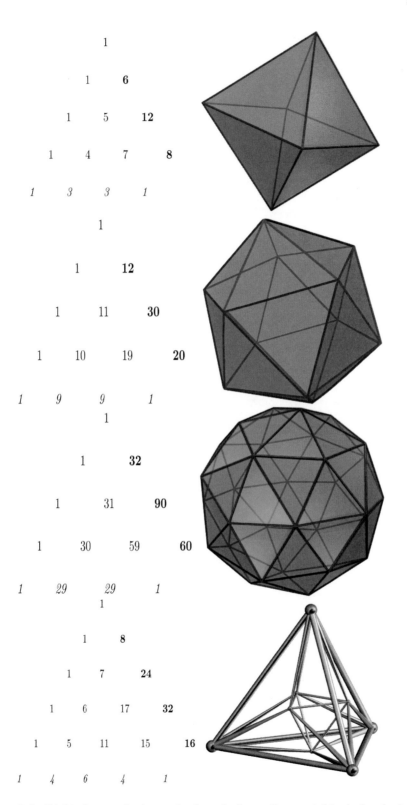

```
            1
        1       6
      1     5       12
    1     4     7       8
  1     3     3     1

            1
        1       12
      1     11      30
    1     10      19      20
  1     9     9     1
            1
        1       32
      1     31      90
    1     30      59      60
  1     29      29      1
            1
        1       8
      1     7       24
    1     6     17      32
  1     5     11      15      16
  1     4     6     4     1
```

Figure 8.4: *Right* the octahedron, the icosahedron, the pentakis dodecahedron and the 16-cell; *left*, Stanley's trick performed on each of them with the f-vectors in bold and the h-vectors at the bottom in italics.

of consecutive entries in the h-vector from the left up to the middle, e.g. the g-vector of the 16-cell is $(3, 2)$, and we define a *multicomplex* Δ to be a collection of vectors with non-negative integer entries with the property that if $v = (v_0, v_1, ..., v_n)$ is in Δ and there exists some other vector $w = (w_0, w_1, ..., w_n)$ such that $w_i \leq v_i$ for each ith entry, then w is in Δ - the *degree* of a vector in Δ is the sum of its entries.

So, in 1971, after a flurry of new results about simplicial polytopes and simplicial spheres, the mathematician PETER MCMULLEN wanted a complete characterisation of the possible f-vectors of simplicial polytopes and he suggested this:

The following conditions are equivalent for a vector v:

1. v *is the h-vector of a simplicial d-dimensional polytope.*

2. v *is the h-vector of a $(d-1)$-dimensional simplicial sphere.*

3. v *is palindromic and there exists a multicomplex Δ such that the ith term of v's associated g-vector, i.e. the vector whose entries consist of the consecutive differences of v's entries up to the middle term, is the number of vectors in Δ with degree i.*

This conjecture was open from that year until Christmas Day of last year, when KARIM ADIPRASITO proved it in its entirety in a paper entitled **Combinatorial Lefschetz Theorems Beyond Positivity**. Parts 1 and 3 of the conjecture were proven in the 20th century and this result was called the g-theorem but it turns out that simplicial spheres are much more general than simplicial polytopes and proving it for the sphere case was much harder.

Why would anyone be interested at all in this conjecture? What we've just seen was an extremely hurried attempt to explain all the material that you need to know to understand just the *statement* of the g-conjecture; many major open problems in combinatorics and other fields today will have a lot of theory around them that one needs to digest but to those for whom the ideas are well-known, the problems can seem very natural. For those who study the combinatorics and structure of polytopes and simplicial complices, it seemed a natural problem to come up with some way of easily determining the different numbers of facets a simplicial sphere can have in each dimension and the above conjecture, now that it's proved, provides such a method (there will only be a finite number of possibly relevant multicomplices and they can checked efficiently on a computer) and, while it seems far away from the simplicity of Euler's polyhedral formula, I hope that the way we explained things that many of the discoveries and investigations seemed natural along the way.

To end off with, I'll provide a list of some of my favourite open problems in combinatorics, i.e. those for which there is no known solution. Some of them are just as complex as the g-conjecture, while others, I hope you'll appreciate, are beautifully simple to state; none of them, however, are simple to solve otherwise someone would have done so.

- Erdos and Turan intially conjectured an even more general form of Szemeredi's Theorem: if one has a subset of $\mathbb{N}$ with the property that the sum of the reciprocals of the numbers in the set does not converge to any finite value, but instead goes off to infinity, then the subset must contain infinitely many k-term arithmetic progressions for every $k \geq 3$. Remarkably, nobody has solved this conjecture even for the special case $k = 3$...

- RONALD GRAHAM, of *Graham's number*, has offered a plane-analogue of the previous conjecture: for any set of points on the plane with integer coordinates and the with the property that the sum of the reciprocals of the squares of their distances from the origin similarly goes off to infinity, then the set must contain infinitely many $k \times k$ square-grids for every $k \geq 3$. He has offered \$1000 for the proof or disproof of this conjecture.

- The Van der Waerden number $W(k)$ is defined to be the least N such that any 2-colouring of the set of numbers from 1 to N must contain a monochromatic arithmetic progression of length k. The problem is to determine roughly how fast $W(k)$ grows, e.g. is it true that $W(k) \leq 2^{k^2}$? We know that $W(k) \leq 2^{2^{2^{2^{2^{k+9}}}}}$, which isn't really that useful.

- Suppose we have an infinite increasing sequence S of natural numbers with the property that no k-segment, i.e. no series of k consecutive elements from S, contains a 3-term arithmetic progression. How big can the density of S be?

- Let the *diameter* of a graph be the length, in edges, of the longest possible distance between any two points in the graph (where the *distance*, to be clear, between two points is the length of the shortest possible path between them). The *Polynomial Hirsch Conjecture* states that if G is the graph of some d-dimensional polytope which has n facets in total, then the diameter of G is bounded above by some polynomial in n and d.[8]

- For any 4-polytope, is it true that $\frac{f_1+f_2}{f_0+f_3}$ is at most some specific real number, r? The belief that this is correct is humorously called the *fat 4-polytope conjecture*.

- *The Hadwiger-Nelson problem* is to determine the number of colours required to colour the graph G, defined as the graph which has a vertex for each point on the plane and two vertices have an edge between them if their corresponding points are distance 1 apart, so that no two vertices joined by an edge are the same colour. We knew that this number was one of $4, 5, 6$ or 7 for many many years but last year an amateur mathematician, AUBREY DE GREY, announced a proof that it could not be 4, so the quest continues...

---

[8] The original Hirsch conjecture stated that every such graph had diameter at most $n - d$, but a 43-dimensional counter example was found!

We hope you enjoyed this issue of ARM and perhaps will investigate some of the ideas mentioned in this article. As always, we end with two challenges:

1. Look up the 4-simplex and draw its face-lattice.

2. The construction in the proof of Schur's theorem allows for many generalisations using Ramsey numbers. Can you come up with any?

Last Note

So there finishes our journey for now - 100 pages of mathematics completed. Yet in mathematics, there's always much more to be said - I would have loved to discuss the wonders of Diophantine approximation, the bizarre quandaries of nonstandard models of arithmetic, objects that are non-integer-dimensional, group theory and its mysterious 196,882-dimensional object with deep connections to the complex numbers. There's much much more that I couldn't even begin to describe, because I don't know about it yet - and that excites me most of all!

Therefore, for those who want some further reading beyond just what was mentioned in each of the chapters, consider the images at the beginning of each chapter, which are relevant to but not necessarily discussed during each text. Here is a list of their names without explanation, so that you can have the joy of discovering more about them:

- Chapter 1: A computer-aided drawing of single members from each of the classes of tesselating pentagon.

- Chapter 2: An image of "Sir Robin", the first elementary knightship discovered in Conway's Game of Life.

- Chapter 3: Images of the first few stages of the geometric figures defining the "gull-wing" variant of the *toothpick sequence*.

- Chapter 4: An illustration by Aldwin Li showing a staircase up many of the smaller countable infinite ordinals.

- Chapter 5: A tiling of the hyperbolic plane in which 4 regular hexagons (here, chopped into 12 black and white sections each) meet at a point.

- Chapter 6: An illustration by Aldwin Li showing a staircase spiralling downwards from ω to some of the larger numbers met in the chapter.

- Chapter 7: An edited image the author took of the Empire State Building while in New York.

- Chapter 8: An image of the maximal (3,3,3) Ramsey Graph, with the coloring showing no monochromatic triangles.

The images on the front cover of this book are, starting from the top-left in clockwise order:

- An image of the Great Dirhombicosidodecahedron.

- The first (ordered by absolute magnitude) 500 Eisenstein primes.

- A projection of the 120-cell.

- The smallest squaring of the square.